Acoustical Impulse Response Functions of Music Performance Halls

Synthesis Lectures on Speech and Audio Processing

Editor
B.H. Juang, *Georgia Tech*

Acoustical Impulse Response Functions of Music Performance Halls
Douglas Frey, Victor Coelho, and Rangaraj M. Rangayyan
2013

DFT-Domain Based Single-Microphone Noise Reduction for Speech Enhancement: A Survey of the State of the Art
Richard C. Hendriks, Timo Gerkmann, and Jesper Jensen
2013

Speech Recognition Algorithms Using Weighted Finite-State Transducers
Takaaki Hori and Atsushi Nakamura
2013

Articulatory Speech Synthesis from the Fluid Dynamics of the Vocal Apparatus
Stephen Levinson, Don Davis, Scot Slimon, and Jun Huang
2012

A Perspective on Single-Channel Frequency-Domain Speech Enhancement
Jacob Benesty and Yiteng Huang
2011

Speech Enhancement in the Karhunen-Loève Expansion Domain
Jacob Benesty, Jingdong Chen, and Yiteng Huang
2011

Sparse Adaptive Filters for Echo Cancellation
Constantin Paleologu, Jacob Benesty, and Silviu Ciochina
2010

Multi-Pitch Estimation
Mads Græsbøll Christensen and Andreas Jakobsson
2009

Discriminative Learning for Speech Recognition: Theory and Practice
Xiaodong He and Li Deng
2008

Latent Semantic Mapping: Principles & Applications
Jerome R. Bellegarda
2007

Dynamic Speech Models: Theory, Algorithms, and Applications
Li Deng
2006

Articulation and Intelligibility
Jont B. Allen
2005

Acoustical Impulse Response Functions of Music Performance Halls

Douglas Frey, Victor Coelho, and Rangaraj M. Rangayyan

ISBN: 78-3-031-01437-6   paperback
ISBN:978-3-031-02565-5   ebook

DOI 10.1007/978-3-031-02565-5

A Publication in the Springer series
SYNTHESIS LECTURES ON SPEECH AND AUDIO PROCESSING
Lecture #12
Series Editor: B.H. Juang, *Georgia Tech*
Series ISSN
Synthesis Lectures on Speech and Audio Processing
Print 1932-121X   Electronic 1932-1678

Acoustical Impulse Response Functions of Music Performance Halls

Douglas Frey
University of Calgary, Calgary, AB, Canada

Victor Coelho
Boston University, Boston, MA

Rangaraj M. Rangayyan
University of Calgary, Calgary, AB, Canada

SYNTHESIS LECTURES ON SPEECH AND AUDIO PROCESSING #12

ABSTRACT

Digital measurement of the analog acoustical parameters of a music performance hall is difficult. The aim of such work is to create a digital acoustical derivation that is an accurate numerical representation of the complex analog characteristics of the hall. The present study describes the exponential sine sweep (ESS) measurement process in the derivation of an acoustical impulse response function (AIRF) of three music performance halls in Canada. It examines specific difficulties of the process, such as preventing the external effects of the measurement transducers from corrupting the derivation, and provides solutions, such as the use of filtering techniques in order to remove such unwanted effects. In addition, the book presents a novel method of numerical verification through mean-squared error (MSE) analysis in order to determine how accurately the derived AIRF represents the acoustical behavior of the actual hall.

KEYWORDS

acoustical impulse response, acoustics, convolution, deconvolution, digital signal processing, exponential sine sweep, frequency response, inverse filter, loudspeaker, music performance hall, spectral analysis, Wiener filter

To my mother Cecelia Elizabeth and my dad John Lorne,
the finest people, and parents, ever.

– Douglas Frey

Contents

Preface

The processing of analog and digital signals is a complex and difficult process in the measurement and documentation of the analog characteristics of a music performance hall. This book explores the methodologies involved in the derivation of acoustical impulse response functions of music performance halls. The importance of software resolution, inverse filters, and digital numerical verification of the final acoustical derivation are example topics that are presented. Accordingly, it is also important to monitor, maintain, and document methodological integrity throughout the process.

The material presented is primarily concerned with concepts in audio, acoustics, and related forms of digital signal processing. Therefore, the book is suitable for a typical undergraduate student in engineering, computer science, or physics. Specifically, one chapter of the book is dedicated to the algorithm design and development of digital filters for convolution and deconvolution in which the numerical process of converting signals from the time domain to the frequency domain is presented. Efficiency of process is examined in a study of the Fast Fourier Transform or FFT.

The methods and procedures presented in this book provide an improved awareness of measurement issues that could lead to an increased level of accuracy in the characterization of the acoustical parameters of music performance halls. A primary function of the results obtained in such measurements is the acoustical preservation of culturally sensitive and important halls. Furthermore, due to the advantages of digital storage and recall, the documentation of the acoustical properties of historic architecture provides the opportunity to understand and study their characteristics for future research and development.

Douglas Frey, Victor Coelho, and Rangaraj M. Rangayyan
March 2013

Acknowledgments

In the spirit of acoustical and academic research, the music performance halls and the Meyer MTS-4 measurement loudspeaker system were provided by the production staff at University Theatre Services, University of Calgary. We thank them for their cooperation.

Advice, encouragement, inspiration, software assistance, and instruction (La Casa Della Musica) have been provided by Dr. Angelo Farina, University of Parma, Italy. We express our respect and gratitude to him.

Methodology and advice on measuring reverberation time (RT60), and the corresponding data for the Eckhardt-Gramatté Hall, were provided by acoustical engineer Niels V. Jordan, of Jordan Akustik, Roskilde, Denmark. We express our thanks to him.

We thank IEEE for providing royalty-free permission to reproduce in the book material from our publications:

D. Frey, V. Coelho, and R. M. Rangayyan, "Spectral Verification of an Experimentally Derived Acoustical Impulse Response Function of a Music Performance Hall," *Proceedings of the 23rd IEEE Canadian Conference on Electrical and Computer Engineering (CCECE2010)*, Calgary, Alberta, Canada, May 2010, pp. 1–4.

D. Frey, V. Coelho, and R.M. Rangayyan, "Filtering and Removal of the Effects of the Transducers on the Acoustical Impulse Response of Concert Halls," *Proceedings of the 22nd IEEE Canadian Conference on Electrical and Computer Engineering (CCECE 2009)*, St. John's, NF, Canada, May 2009, pp. 368–371.

D. Frey, V. Coelho, and R.M. Rangayyan, "The Loudspeaker as a Measurement Sweep Generator for the Derivation of the Acoustical Impulse Response of a Concert Hall," *Proceedings of the 21st IEEE Canadian Conference on Electrical and Computer Engineering (CCECE 2008)*, Niagara Falls, ON, Canada, May 2008, pp. 301–304.

Douglas Frey, Victor Coelho, and Rangaraj M. Rangayyan
March 2013

List of Symbols, Abbreviations, and Nomenclature

A/D	Analog-to-Digital
AES	Audio Engineering Society
AIRF	Acoustical Impulse Response Function
BW	Bandwidth
CD	Constant Directivity
DAT	Digital Audio Tape
dB	decibels
D/A	Digital-to-Analog
DFT	Discrete-time Fourier transform
DSP	Digital Signal Processing
DWT	Discrete wavelet transform
ESS	Exponential Sine Sweep
f	frequency in Hertz
FFT	fast Fourier transform
FT	Fourier transform
GUI	Graphical User Interface
HF	High Frequency
Hz	Hertz
IEC	International Electrotechnical Commission
IFFT	Inverse fast Fourier transform
ISD	Integrated Spectral Difference
ISO	International Standards Organization

k	wave number $2\pi/\lambda$
λ	wavelength υ/f in meters
LF	Low Frequency
LFSR	Linear Feedback Shift Register
LPF	Low Pass Filter
LTI	Linear Time Invariant
MF	Mid Frequency
MLS	Maximum Length Sequence
MMSE	Minimum Mean Squared Error
MSE	Mean Squared Error
η	efficiency
ω	angular frequency in radians per second
ORTF	Office de Radiodiffusion Television Francaise
π	pi radians
PSD	Power Spectral Density
RMS	Root-Mean-Squared
RT60	Reverberation Time
RTA	Real-Time Analysis
SIMD	Single-Instruction Multiple-Data
SNR	Signal-to-Noise Ratio
SPL	Sound Pressure Level
TSIF	Transducer System Inverse Filter
υ	velocity of sound in meters/second

And if the mountain should crumble

Or disappear into the sea

Not a tear, no not I

Stay in this time

Stay tonight in a lie

Ever after

This love in time

And if you save your love

Save it all

Bono of *U2*; from *The Unforgettable Fire* (1984)

CHAPTER 1

Introduction

Acoustic characterization of music performance halls has been an active area of research since the 1970s. In 1975, Gerzon [1] presented the concept of systematically collecting the (3-D) impulse response measurements of ancient theatres and concert halls in order to preserve their acoustical characteristics for posterity and provide the opportunity for recovery of the original, acoustic sound field by a future technology. As computationally intensive operations such as convolution and deconvolution became increasingly possible through developments in digital signal processing (DSP), measuring an acoustical impulse response function (AIRF) of a music performance hall became more practical and efficient. A major contribution to the field of acoustic measurement has been the development of the exponential sine sweep (ESS) acoustic measurement process, attributed to Farina [2], at the University of Parma, Italy. Since 2000, audio scientists in Europe, and particularly in Italy, have developed large-scale archival projects where the goal is to record and preserve the acoustical characteristics of their historic concert halls and opera houses [3], long known to contemporary performers and historians as optimal performance spaces.

1.1 ACOUSTIC MEASUREMENT OF MUSIC PERFORMANCE HALLS IN CANADA

Comparatively, the characterization of music performance halls through the measurement of various acoustic parameters is a new practice in western Canada. Nationally, there have been important studies such as those by Bradley [4]. Absent in the existing research of Canadian concert halls are those in western cities such as Vancouver, Calgary, and Edmonton, whose acoustic characteristics would be valuable to include in the developing database devoted to the acoustical documentation and preservation of important facilities. One example is the original Jubilee Auditorium in Edmonton, which ranked in the top 100 finest sounding concert halls in the world [5]. Unfortunately, it no longer exists in its original form as it was closed for an entire year for renovations and reopened in 2005. Another example is the Rozsa Centre Complex at the University of Calgary which houses three music performance halls: the Eckhardt-Gramatté Chamber Music Hall, the Husky Oil Great Hall, and the Rehearsal Hall. Documenting the acoustical characteristics of these spaces not only ensures a computer or numerical archive for immediate recall and analysis, but can also provide a visual and aural interpretation of the quantitative nature and propagation of music and sound in the facility. In this respect, it is possible to examine the mathematical and psychophysical effects, both positive and negative, of the architectural acoustics of a music performance hall. In the present book, through novel developments in acoustic characterization and using the ESS measurement technique as a basic framework, AIRFs of the three music performance halls at the Rozsa Centre are measured, analyzed, and documented.

1.2   GEOMETRICAL ACOUSTICS

The different orders of reflections that together form a characteristic analog or continuous-time AIRF of a hall can be theoretically mapped beforehand according to a concept drawn from geometrical acoustics known as the "phantom source" [6]. A useful feature of this method is that it is possible to determine geometrically the paths of the characteristic orders of reflections independent of the location of the listener. That is, the calculation is based upon the position of the sound source and the location of each wall off which it reflects. Consequently, distances and delay times can be determined as well. Physical modeling through the use of phantom sources is not used as part of the present book but can be important in the derivation of a geometrically based computer model or digital simulation of acoustic space [6].

1.3   THE IMPORTANCE OF MEASUREMENT ACCURACY

The derivation of an accurate and representative AIRF is difficult. Fundamentally, the propagation and modification of sound waves in an acoustic space are physically and mathematically complex phenomena. The multitude of time-delayed interferences, both constructive and destructive, of the various orders of characteristic reflections along with the different harmonic orders of the original audio signal, produce acoustical patterns that are difficult to quantify and characterize. Subjective aural qualification is not sufficient by itself, as for an accurate representation, the primary goal is objective numerical and visual quantification.

To complicate matters, difficulties of accuracy in the measurement process itself are caused by the limitations associated with the hardware and software of the available measurement tools. With the ESS measurement process, hardware tools include the necessary loudspeaker and microphone transducers involved in the initial stage of producing and recording the measurement sweep in the hall. Software tools include the different DSP modules used throughout the entire process, especially those used in post-production data processing. Therefore, in order to address these limitations [7], it is important to build a library of modular subsystem measurement techniques, such as the development of inverse filters for specific applications, or the use of spectral analysis techniques in order to monitor signal uniformity throughout the entire measurement process. An example is the isolation and measurement of the frequency response of the reference or measurement loudspeaker in order to eliminate, or at least, limit its effect on the final derived AIRF. The performance of such techniques ensures that measurement accuracy can be regulated and maintained in order to provide a faithful and representative AIRF derivation.

1.3.1   SOFTWARE ISSUES IN DEVELOPING AN ACCURATE REPRESENTATION

Developments in DSP have enabled the theoretical research and conceptual proposals of audio scientists such as Gerzon [1] to become practical realizations. However, the increasing efficiency and popularity of DSP is a "double-edged sword," as the commercial audio software market, including internet freeware and shareware, has been saturated with widely accessible but minimally regulated audio recording and measurement products, making it difficult to separate and eliminate inferior software designs that produce excessive amounts of distortion in a signal. As an example, there are various brands of commercially available graphic equalizer modules, that all perform the same function, but are produced by different manufacturers. Furthermore, these various modules have identical user interfaces and parameter control, indicating that they will perform similarly as well. Upon further analysis however, they are actually quite different in their processing algorithms. By taking the time to perform spectral analysis of specific processing modules, it is possible to identify problems of inconsistency and potential inaccuracies in software performance [7]. Consequently, on a macro level, creating a database of modular subsystem measurement techniques could facilitate the development of a representative AIRF for a given hall or environment.

1.3.2   VERIFICATION OF THE REPRESENTATION

Once the derivation of the AIRF has been accurately processed, the final step is the verification of the measurement. This is essentially an examination of how accurately the derived AIRF represents the actual hall it was measured in. It is important to note that numerical and visual verification is the goal, as aural verification through listening tests is a subjective process and does not provide objective accuracy. Furthermore, some kind of numerical verification is necessary for absolute comparison, and should actually be a specification requirement with commercially available AIRFs. In the existing world of marketable software, a manufacturer may claim that their AIRF is representative of a certain hall, such as the Royal Opera House in London, without providing numerical data to substantiate the claim. In an effort to create a standard for objective representation, the present book provides a novel method for AIRF spectral verification, and furthermore, suggests approaches to improve on the proposed method for further study.

CHAPTER 2

A Review of Acoustic Measurement Techniques

One traditional method for measuring the AIRF of a music performance hall is to generate a brief, omnidirectional impulse of finite amplitude. These types of signal sources are typically produced by a starter's pistol or through popping a balloon [10]. The impulse response of the hall is recorded on either a digital audio tape (DAT) machine or a computer (with appropriate software) for subsequent processing and analysis. Impulsive sources such as a starter's pistol or balloon might provide a usable recorded signal-to-noise ratio (SNR) but there are problems of stability, repeatability, and bandwidth of the frequency response. Consequently, there is the absence of a normalized or consistent spectrum for the test signal, which makes it difficult to determine both the absolute spectral response of the hall being measured and the absolute sound pressure level (SPL) of the source.

Another traditional, commonly used acoustical measurement technique is called the Maximum Length Sequence (MLS) and is based upon using a loudspeaker-generated wideband, deterministic, pseudorandom test signal as a measurement source [2]. MLS is a type of pseudorandom binary sequence generated by an Nth-order, linear feedback shift register (LFSR) having a length or period of $2^N - 1$, where N corresponds to the number of shift registers or bits in the system [11]. Instead of following a random pattern of objective unpredictability, the MLS signal is pseudorandom because it consists of a deterministic sequence of pulses that repeats itself in a periodic manner. An MLS test signal is typically pseudorandom white noise (see Figure 2.1). Parameters such as the number of shift registers and sampling (clock) frequency are important in order to avoid problems such as time-aliasing error (due to circular convolution or deconvolution as opposed to linear). Aliasing occurs if the period $(2^N - 1)$ of the repeated input signal is shorter than the duration of the impulse response of the device or room [2]. Furthermore, the requirement of time-invariance with the MLS process demands that the excitation signal be tightly synchronized with the digital sampler employed for recording the system's response [2]. As illustrated in Figure 2.2, the MLS spectrum is reasonably flat, offering a "uniform" measurement reference source.

As mentioned in the previous paragraph, there are limitations in using the MLS technique for the derivation of an AIRF. First, as it is based on calculating the cross-correlation function of deterministic sequences; it is useful only for a time-invariant system. Second, the measured nonlinear harmonic distortion of the system (loudspeakers) is spread throughout the derived impulse response [2]. This makes it impossible to isolate and remove these distortion products from the measured linear response. In this manner, the MLS technique is not tolerant to any nonlinearity in

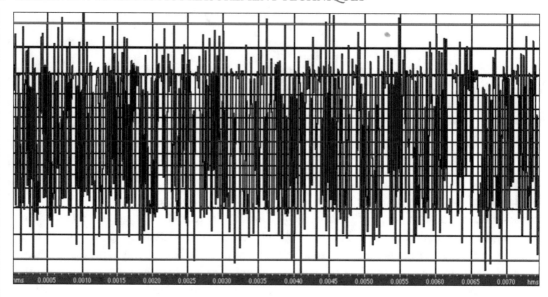

Figure 2.1: Time-domain representation of the MLS test signal. Time-scale: 0.5 ms per division; amplitude-scale: 3 dB per division.

the measurement system. Finally, with MLS, in order to reduce measurement error due to additive system noise, the SNR of the AIRF can be improved only by averaging together a number of recorded sequence repetitions prior to the cross-correlation of the original test signal and the recording of the hall's response. In this case, multiple averaging becomes a questionable technique as, aside from the issue of time variation of waveform propagation in a music performance hall, it also creates the necessity of absolute synchronization among the repeated sequences in order to produce an average that is accurate.

2.1   THE EXPONENTIAL SINE SWEEP (ESS) TECHNIQUE

At the Audio Engineering Society (AES) Paris Conference in 2000, Farina presented a paper on the successful employment of the ESS technique [2] for measuring acoustical parameters. Due to the advantages it presents over previous techniques such as MLS, the employment of a loudspeaker-generated ESS measurement signal has become increasingly common in acoustic measurement, especially for measurements of acoustic impulse responses and loudspeaker harmonic distortion [2]. The developments and practicality of the technique have been possible due to the increase in the efficiency of DSP which has allowed computationally intensive operations such as convolution of lengthy sequences of signal data to be performed in real time.

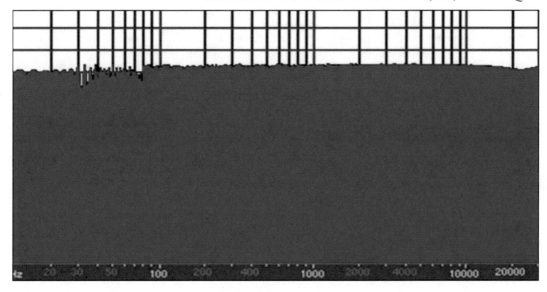

Figure 2.2: Frequency-domain representation of the MLS test signal. Frequency-scale: 0 Hz to 30 kHz (log scale); amplitude-scale: 12 dB per division.

The ESS approach is similar to the methods previously discussed in that, hypothetically, the acoustics of a hall can be modeled as a linear time-invariant (LTI) system, and described as a single-input, single-output, "black box" (see Figure 2.3). In reality, the electro-acoustic behaviour of the measurement loudspeaker contributes to system nonlinearity and the subsequent waveform propagation in the theatre is not perfectly time-invariant. However, the ESS technique of acoustical measurement lessens the impact of these issues to the point where their impact is minimal. This is discussed in the next Section 2.1.1.

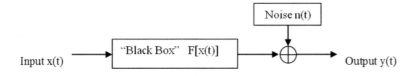

Figure 2.3: Acoustics of a hall modeled as an LTI system.

There are essentially two independent processes necessary in order to determine the ESS derivation of an AIRF of a music performance hall. First, the loudspeaker-generated measurement ESS is recorded at a specific location in the hall (see Figure 2.4).

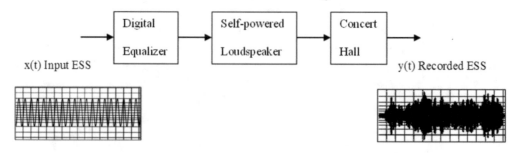

Figure 2.4: Process of recording the ESS in a concert hall.

A computer generates and records the ESS which is initially sent through a loudspeaker prefilter (digital equalizer) before being sent to the measurement loudspeaker on stage. The measurement ESS can be defined mathematically as [2, 8],

$$x(t) = \sin[(K) \cdot (e^{\{(t/T) \cdot (L)\}} - 1)]$$

where

$$K = [(\omega_1 T)/L]$$

where ω_1 is the starting angular frequency,
ω_2 is the ending angular frequency,
T = the total duration of ESS in seconds, and
$L = [(\ln |\omega_2/\omega_1|)]$.

The second process involves the post-processing of data and the derivation of the AIRF, through the deconvolution of the recorded ESS and the previously calculated inverse filter of the original input ESS (see Figure 2.5).

On a logarithmic scale, the instantaneous frequency response of an ESS (see Figure 2.6) is similar to the decreasing −3 dB per octave sloping behavior of a pink-noise filter. In contrast, white noise has an increasing +3 dB per octave sloping characteristic as it contains equal energy per Hz [12]. Consequently, pink noise is white noise that has been modified with a pink-noise filter [12, 13]. Therefore, the spectrum of the resulting pink noise is flat or uniform as it generates an equal amount of energy per unit octave and not per unit Hz [12, 14].

Accordingly, an ESS measurement signal with a range of 22 Hz to 22 kHz will sweep slowly at the low frequencies, where it possesses its greatest amount of energy, and then sweep progressively faster as the frequency increases, according to the logarithmic scale. The decreasing −3 dB slope is important as the measured ESS inverse filter needed for deconvolution (see Figure 2.5) must then compensate for the decreasing slope of the original sweep by exhibiting a +3 dB per octave increase over the same bandwidth (see Figure 2.7).

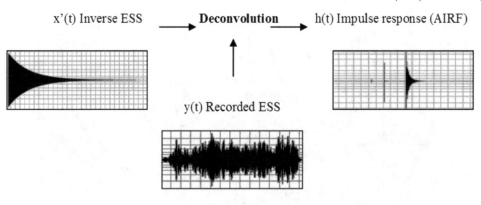

Figure 2.5: Post-processing derivation of the AIRF.

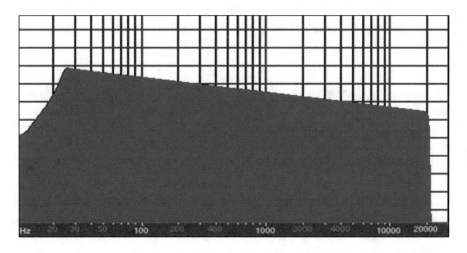

Figure 2.6: Spectrum of the original 22 Hz to 22 kHz ESS. Frequency-scale: 0 Hz to 30 kHz (log scale); amplitude-scale: 12 dB per division.

2.1.1   ADVANTAGES OF THE ESS TECHNIQUE

The ESS technique of acoustic measurement offers certain advantages over others such as MLS. The ESS technique does not limit the measurement being performed to an LTI environment as is required with MLS [2]. This is because first, in addressing the issue of measurement nonlinearity, mainly due to the measurement loudspeaker, using an ESS as a measurement sweep (as opposed to a linear sweep) allows for the isolation and removal of the loudspeaker-generated, nonlinear harmonic distortion products. That is, the association of the ESS inverse filter, and the process of linear deconvolution (as opposed to circular) [2], ultimately enables the separation and isolation of

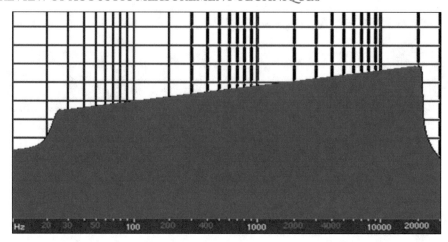

Figure 2.7: Spectrum of the ESS inverse filter. Frequency-scale: 0 Hz to 30 kHz (log scale); amplitude-scale: 12 dB per division.

each harmonic response in that they are positioned in sequence either prior to or after the primary linear response of the hall [15]. In this manner, the increasing orders of harmonic distortion can be identified and isolated for analysis as each harmonic order is individually represented by its own characteristic impulse response. Figure 2.8 illustrates the ESS deconvolution output for the Eckhardt-Gramatté Hall produced by the "linear convolution with clipboard" module developed by Farina [16]. In this example, the nonlinear harmonic responses are positioned in a sequence of increasing order (right to left) prior to the linear response, which is the result of using a rising frequency ESS as a measurement signal. A falling frequency ESS will result in the harmonic products being positioned to the right of the linear response [15]. Either way, due to this characteristic result of using the ESS as a measurement source, the nonlinear harmonic distortion products generated by the measurement loudspeaker do not pose a problem as they may be easily removed.

Figure 2.9 illustrates the spectrogram for the recorded ESS from the Eckhardt-Gramatté Hall and provides greater insight to the deconvolution result of Figure 2.8. Due to the nature of the ESS, as opposed to a linear sweep, the increasing orders of nonlinear harmonic distortion appear as straight lines, with identical linear slopes, above the fundamental excitation sweep. According to Farina, linear deconvolution with the ESS inverse filter causes the recorded ESS plot of Figure 2.9 to "stretch" and rotate counter-clockwise, so that the linear and nonlinear responses are represented as clearly separated vertical lines or independent impulse responses [2] (see Figure 2.8). (The distinction between linear and circular convolution [and deconvolution] is discussed in Chapter 4.)

In addressing the issue of time variation, another advantage of using the ESS as a measurement signal is that its power is distributed linearly over the entire duration of the sweep. This is illustrated by the decreasing slope (−3 dB per octave) of the ESS spectrum in Figure 2.6. In the post-processing

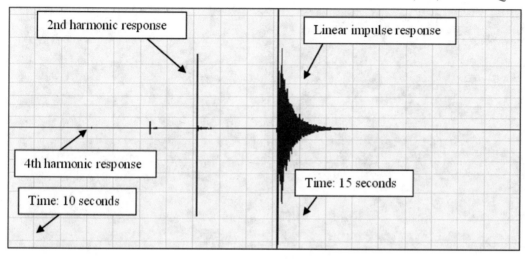

Figure 2.8: Output of the convolution (deconvolution) module developed by Farina using the recorded 15-second ESS from the Eckhardt-Gramatté Hall and the inverse filter of the original ESS. Prior to the linear response are the second, third, and fourth orders of harmonic distortion. Time-scale: 0.5 s per division; amplitude scale: 3 dB per division. Reproduced, with kind permission from IEEE, from D. Frey, V. Coelho, and R.M. Rangayyan, "The loudspeaker as a measurement sweep generator for the derivation of the acoustical impulse response of a concert hall," *IEEE Canadian Conference on Electrical and Computer Engineering (CCECE)*, Niagara Falls, ON, Canada, 4–7 May, 2008, pp. 301–304. © IEEE.

stage of deconvolution, the amount of signal power is maintained but compressed back into a short response, resulting in an SNR improvement of 60 dB or more in comparison with the generation of a single impulse having the same maximum amplitude [2]. Therefore, synchronous averaging of multiple responses is not necessary in order to produce a good SNR, and consequently, the estimated response is not affected by any time variation as a single recorded sweep is sufficient for the derivation of the AIRF. It is important to note that a linear sine sweep produces a good SNR as well, as its power distribution is similar to that of the ESS in Figure 2.6, but instead of a linearly decreasing slope of −3 dB per octave, a linear sweep produces a constant and uniform slope following the behavior shown in Figure 2.2. However, the deconvolution of a linear sweep and its corresponding inverse filter does not separate and isolate the nonlinear harmonic responses of the measurement system [15].

2.1.2   PERFORMANCE OF ELECTRO-ACOUSTIC TRANSDUCERS

Developments in DSP have improved the efficiency of the ESS measurement process and made it a practical alternative in acoustic characterization of devices and halls. The weak link in the process

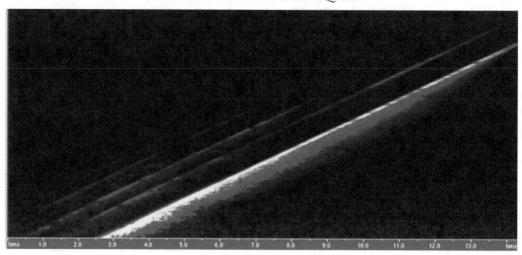

Figure 2.9: Spectrogram of the recorded 15-second ESS performed in the Eckhardt-Gramatté Hall. Above the excitation sweep (brightest trace) are the second, third, and fourth orders of nonlinear harmonic distortion shown in Figure 2.8. Time-scale: 1.0 s per division; frequency-scale: 100 Hz to 30 kHz (log scale).

is still the analog hardware of the transducer system, namely the loudspeakers and microphones [7]. The transduction of sound signals from one form to another (electro-mechanical-acoustic) is mathematically complex and difficult to quantify as physics-related problems such as impedance matching (from one form to another) and waveform phase distortion make it difficult to design analog transducers that are efficient. Chapter 3 presents the problems of loudspeaker efficiency and performance including specific parameters that need to be examined in determining the proper type of loudspeaker to be used as a measurement sweep generator in the ESS derivation of the AIRF of a music performance hall.

Consistency is another problem with transducer hardware as the actual measured field performance of a loudspeaker or microphone may differ significantly from the same "ideal" parameter specifications published by the manufacturer. As an example, a microphone under field measurement in an anechoic chamber may, in fact, produce a different polar response in comparison to the manufacturer's published specifications [8]. Furthermore, this problem of polar response distortion may be a function of frequency, making the problem itself inconsistent and more difficult to manage.

2.2   SOFTWARE LIMITATIONS IN MEASUREMENT PROCESSING ACCURACY

In an attempt to maintain processing accuracy and consistency throughout the derivation of an AIRF, in the course of the present work, it was discovered that it is useful to perform spectral analysis routinely on particular software modules in order to see how they may produce inaccuracies in their processing of an input signal. Software modules with specific audio processing functions, such as graphic equalization, from different manufacturers but with identical graphical user interfaces (GUI), may actually be very different in behavior. As an example, for the present work, separate spectral pink noise analysis was performed on two different digital one-third-octave graphic equalizer modules, with identical interfaces, to determine if there were any differences or potential inaccuracies in algorithm performance.

The result in Figure 2.10 was obtained using the Adobe Audition 2.0 one-third-octave graphic equalizer module and provides a reasonable, constant, linear consistency to the corresponding fader settings of 20 dB attenuation in the frequency band of 2.5 kHz to 20 kHz. The result in Figure 2.11 is

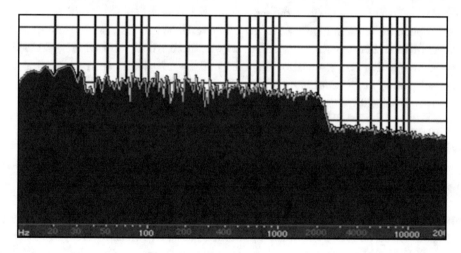

Figure 2.10: Pink-noise filter spectral analysis of Adobe Audition 2.0 one-third-octave graphic equalizer module, with constant 20 dB attenuation from 2.5 kHz to 20 kHz. Frequency-scale: 0 to 20 kHz (log scale); amplitude-scale: 12 dB per division. Reproduced, with kind permission from IEEE, from D. Frey, V. Coelho, and R.M. Rangayyan, "Filtering and removal of the effects of the transducers on the acoustical impulse response of concert halls," *IEEE Canadian Conference on Electrical and Computer Engineering (CCECE)*, St.John's, Newfoundland and Labrador, Canada, 3–6 May, 2009, pp. 368–371. © IEEE.

from a similar graphic equalizer module produced by another manufacturer, using identical settings of 20 dB attenuation, in the same frequency band. However, Figure 2.11 indicates a nonlinear slope where the amount of attenuation is not constant according to the set amount of 20 dB. This does not

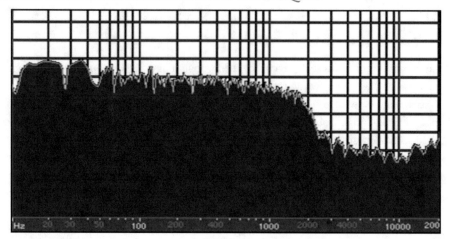

Figure 2.11: Pink-noise filter spectral analysis of a similar one-third-octave graphic equalizer module produced by another manufacturer, with a different result, but identical equalizer settings as the Adobe module used to obtain the result in Figure 2.10. Frequency-scale: 0 Hz to 20 kHz (log scale); amplitude-scale: 12 dB per division. Reproduced, with kind permission from IEEE, from D. Frey, V. Coelho, and R.M. Rangayyan, "Filtering and removal of the effects of the transducers on the acoustical impulse response of concert halls," *IEEE Canadian Conference on Electrical and Computer Engineering (CCECE)*, St.John's, Newfoundland and Labrador, Canada, 3–6 May, 2009, pp. 368–371. © IEEE.

necessarily indicate an inferior software module, but it does indicate a different processing algorithm. More importantly, it illustrates an inconsistency in that its output does not accurately represent the parameter settings of its control panel (interface).

Therefore, it is important to note that a given DSP software module may produce different results than expected according to the particular algorithm driving the program. If this is of concern, it is useful to perform spectral analysis of the software in question, and if required, replace it with a module from a different manufacturer. In this manner, measurement accuracy can be regulated and maintained.

2.3   REMARKS

Even with the positive research and developments in both DSP and analog hardware, the process of acoustic measurement and characterization is still difficult due to the many preliminary tests, analyses, and other procedures that are necessary to ensure reliable and reproducible measurements. Proper objective quantification is the ultimate goal. Because preliminary testing procedures can be time consuming, it is important to develop a modular library of monitoring and alignment techniques that can be used quickly throughout the entire process. Process regulation in itself is difficult but necessary in order to produce a representative AIRF derivation that is identical to the acoustical impulse response of the actual hall. A primary concern of the present work is to develop proper analytical, filtering, and spectral measurement techniques that help to maintain process integrity. Chapter 3 presents an analysis of the measurement source and a discussion of the problems of specific loudspeaker parameters and performance issues that need to be examined in order to determine which type of loudspeaker should be used as a measurement sweep generator in the derivation of an AIRF of a music performance hall.

CHAPTER 3

The Loudspeaker as a Measurement Sweep Generator

To measure an AIRF of a music performance hall in an objective and accurate manner, using the ESS process [2, 8], the measurement engineer needs to determine the requirements of the stimulus being generated by the measurement loudspeaker on stage. As an example, an AIRF derivation may be part of a comprehensive document where various acoustical parameters have been measured [17] according to the methodology presented in the ISO 3382-1 standard [18] on the acoustical measurement of performance spaces. However, upon further analysis, some of the required measurement parameters that are part of this methodology have specific limitations, which affect the transparency and accuracy of an acoustical impulse response measurement. Therefore, as an alternative, an AIRF derivation may also be part of an experiment that addresses some of the technical and musical limitations imposed by the requirements of the ISO standard.

The primary issues of concern with the required parameters presented in ISO 3382-1, concerning the derivation of an AIRF of a music performance hall, are the prescribed polar response of the measurement loudspeaker and the resulting or corresponding limitations of frequency response. These issues dramatically restrict the bandwidth of the final AIRF derivation. In the measurement of an AIRF, the ISO 3382-1 standard requires the use of a measurement loudspeaker with an omni-directional polar response (pattern of radiation). Theoretically, there are good reasons for this. First, this requirement is consistent with the traditional standards of industrial building acoustics [14] where the reference source is a loudspeaker with an "evenly" distributed, spherical pattern of radiation. Second, the omnidirectional pattern helps to ensure accuracy and consistency over repeated measurements. Finally, the omnidirectional polar response also helps to ensure transparency of process in that source and receiver positions may be interchanged (when using an omnidirectional microphone) [2, 8]. However, despite these advantages, omnidirectional loudspeakers (dodecahedrons) also have extensive limitations which affect their ability to be used as reference sources in the measurement of music performance spaces.

Certainly, it appears that the ISO 3382-1 standard has been designed with the purpose to accommodate the omnidirectional pattern and frequency response limitations of multiple element and multiple plane (dodecahedron) loudspeaker systems [8]. Therefore, because of these limitations of dodecahedron loudspeakers, it is not surprising that the ISO 3382-1 standard has similar limitations. Furthermore, these limitations are perhaps not such a factor in industrial measurement, where a wide frequency response beyond 4 kHz may not be important, but they may be contradictory to the requirements of music that the performance hall will have been designed to accommodate.

Loudspeaker parameters of directivity and frequency response are directly related to each other. Due to the problems of mid- and high-frequency (HF) directivity associated with direct-radiator elements and those of dodecahedron loudspeakers, the maximum frequency required by the ISO 3382-1 standard is the octave band centered at 4 kHz. For a full-range dodecahedron, this is approximately the frequency where the stated maximum acceptable deviation in directivity (beam-width) of ±6 dB occurs [8]. It is important to note that the empirical definition or accepted beam-width of any loudspeaker represents the angle that is bounded by the points where the sound pressure level (SPL) is 6 dB lower than the on-axis level [12]. Therefore, in the case of a full-range dodecahedron, all frequencies above 5 kHz are essentially unusable because they do not fall within the ±6 dB beam-width of the loudspeaker and the ISO 3382-1 standard. Correspondingly, a single-way dodecahedron system actually ceases to produce an omnidirectional pattern at approximately 1 kHz, where it begins to illustrate an increasingly directional or "beaming" polar response [8]. Consequently, at higher frequencies, the "omnidirectional" pattern of the dodecahedron deviates and becomes inconsistent with the production of several peaks and valleys in its polar response.

These limitations are largely due to the physical nature of the single-way, wide-band, direct-radiator elements associated with multiple-element/plane loudspeaker systems (dodecahedrons). An element with a smaller diameter will illustrate an extended HF response, but a reduced low-frequency (LF) response. And of course, a larger element will produce a reduced HF response but an extended LF response. Consequently, different sizes of dodecahedrons will produce differences in directivity and frequency response [8]. Of course, HF equalization is an option, but the excessive equalization needed to produce a flat spectrum prevents it from being considered an efficient solution. Furthermore, no amount of equalization will stabilize the radiation pattern and make it consistent over a wider frequency range.

Another consideration is the frequency response that the performance hall was designed to reproduce. There are many different musical instruments in a symphony orchestra that have a cumulative frequency response of approximately 20 Hz to 20 kHz. Consequently, in the present work, an important issue is whether or not the measurement loudspeaker should possess a correspondingly similar frequency response. In addition, because a music performance hall may be considered a far-field environment, it is necessary to consider whether or not the measurement loudspeaker should possess certain directional characteristics and be capable of radiating the reference stimulus to the more remote locations of the hall. In this manner, the characteristic order of reflections of the hall may be generated at these distances. Loudspeakers with omnidirectional patterns are designed to be used primarily in near-field applications and are not suitable in far-field environments. In considering these issues, an increasingly faithful representation of the hall may be derived. Furthermore, if an AIRF derivation is to be used for auralization purposes, a loudspeaker with an omnidirectional pattern will not produce satisfactory results [10].

Accordingly, one interesting application of measurement for auralization is the impulse response measurement of opera houses in which the simulation of a vocal performance is a concern. In their study on Italian opera houses, Farina et al. [3] used a high-quality, direct-radiator-type, near-field, DSP studio monitor (with a directional HF dome tweeter element) as the loudspeaker representation of a singer. More recent developments in loudspeaker directional control for this type of application include special dodecahedrons where each diaphragm element may be individually switched on or off and filtered through DSP [19]. Theoretically, this provides sufficient control to simulate the 3-D directional or spatial characteristics of a musical instrument or voice. Once again, it is important to note that both near-field monitors and multi-planar loudspeaker systems such as dodecahedrons employ direct-radiator loudspeaker elements (cone or dome) for signal reproduction. An important disadvantage of direct-radiator loudspeaker elements is that they possess limited directional control as their output directivity pattern is diffuse, unpredictable, and difficult to quantify [20]. Therefore, they are severely limited in their ability to radiate or propagate sound in a far-field environment. To complicate matters, this characteristic is a function of frequency [20].

The levels of element diffusion are increased further due to the immediate interface of the direct-radiator element with the surrounding air mass (impedance mismatch). This produces a corresponding inefficiency in the transfer of acoustical energy from the loudspeaker to its environment. Accordingly, this inconsistency of acoustical impedance also produces refraction (reflection) of sound energy where the output from the loudspeaker is folded back upon itself [20]. This severely affects the transparency or accuracy regarding the output of direct-radiator elements (time-alignment and phase). In addressing these issues, horns and waveguides are designed to match the acoustical impedance of direct-radiator elements with the surrounding air, and therefore suffer much less in terms of refraction (reflection) and loss of acoustical energy. Horns and waveguides are, in effect, more transparent from input to output. Because of this, they also help to concentrate the natural diffusion and energy of a direct-radiator element (compression driver, horn throat) into a narrower, more efficient, and increasingly predictable radiation pattern (horn mouth) [12, 20, 21].

3.1   THE MEASUREMENT LOUDSPEAKER

In the present calculation of an AIRF, it is important to select a measurement loudspeaker that can offer directional and quantifiable radiation characteristics in the far-field environment of a music performance hall. Furthermore, in order to produce an AIRF derivation with a wide bandwidth, the loudspeaker should possess a frequency response of approximately 20 Hz to 20 kHz. For increased accuracy as a measurement source, the loudspeaker should also exhibit proper time-alignment of each element in the system, in order to simulate the phase coherency of a single point-source. It is important to note that the quantification of directional control encourages predictable loudspeaker directivity and accuracy, especially in using the loudspeaker as part of an array, which can dramatically extend the coverage pattern of the measurement source if desired [12].

Therefore, an experimental alternative for a measurement loudspeaker might be a high-quality, DSP time-aligned sound reinforcement loudspeaker system with an HF compression driver and horn, as opposed to an HF direct-radiator cone or dome element [9]. A loudspeaker design using direct-radiator-type components is physically and acoustically different from a loudspeaker incorporating compression driver and horn elements. A horn-loaded loudspeaker is used primarily in live sound reinforcement systems where consistent, directional, and predictable far-field coverage is important. Consequently, if the measurement engineer needs to record an ESS at the twentieth row or in the upper balcony of a concert hall, a dodecahedron will not accurately project its direct sound or properly stimulate and generate the characteristic order of reflections and vibrations of the hall so that they reach the recording device. That is, the earlier reflections at these locations will not only be lower in magnitude but the reflections that occur later in time will not be generated and detected [22]. This is important in the development of an AIRF that represents the hall it was measured in.

There are always compromises in determining which classification of loudspeaker is best suited for a specific application. In the realm of loudspeaker design and performance, a common issue is the trade-off between efficiency and frequency response (or bandwidth). As an example, traditionally, with constant directivity (CD) horn designs, their advantage of maintaining a constant polar pattern, independent of frequency, is achieved at the expense of a flat, wide, on-axis amplitude response with frequency [12, 21]. Due to this, the angle of constant directivity with an HF horn will be maintained over the frequency range of interest but may also suffer from a −3 dB HF roll-off point at about 16 kHz [23]. Despite this compromise in extended HF response, it must be considered that, in the far-field environment of a concert hall, due to its higher level of efficiency (increased acoustical output), directivity control, and compression driver output phase coherence (phasing plug), a CD horn loudspeaker system will project much farther uniformly and properly stimulate the characteristic orders of reflections at the more remote locations of the hall. This is the function of the classification of horn loudspeakers.

The class of direct-radiator loudspeakers and associated physical characteristics, resulting in limited efficiency, directional control, and phase coherence, is discussed in the next section. Horn loudspeakers have their own set of complex design and performance issues, but in measuring the acoustics of a large environment, they offer improvements in efficiency and acoustical output, directivity control, and phase coherency. Compression driver and horn loudspeakers, and their characteristic phasing plug, are presented in Section 3.2.2.

3.2   CLASSIFICATIONS OF LOUDSPEAKERS

3.2.1   DIRECT-RADIATOR LOUDSPEAKERS

Loudspeakers are essentially electromagnetic transducers that convert input electrical energy to output acoustical energy. There are two principal classes of loudspeakers: those in which the vibrating surface, called the diaphragm (usually in the form of a cone or dome), radiates sound directly into the surrounding air; and those in which a horn is interposed between the diaphragm and the air [20]. The former describes the direct-radiator classification of loudspeakers. One disadvantage of loudspeakers of this class is that they characteristically possess low efficiency. Loudspeaker efficiency is effectively defined as the ratio of the acoustic sound power output divided by the electrical power input; that is,

$$\text{Efficiency} = \eta = \frac{\text{output acoustic power}}{\text{input electrical power}}$$

Therefore, the efficiency of any loudspeaker is determined by its ability to maintain energy levels in the transition from electrical to acoustical power. A high level of efficiency is an important attribute in the transparency characteristic of any measurement loudspeaker source—that is, an output that maintains both a frequency and phase response equal to that of the input. Referring to the common, traditional loudspeaker model of a vibrating circular piston mounted in a large or "infinite" baffle, the efficiency of radiation of sound from one side of a well-designed direct-radiator loudspeaker is usually no more than a few percent [20]. Typically, direct-radiator-type high-fidelity (hi-fi) speakers and studio monitors have efficiency values that are between 0.2% and a maximum of 2%. In fact, most monitor systems of this type are, in the mid-band range, no more than about 1% efficient [24]. However, despite the disadvantages of low efficiency, it is important to note that there is no correlation between loudspeaker efficiency and the highly subjective issue of sound quality.

The advantages of direct-radiator loudspeakers are that they may be produced at low cost and in small size, while possessing, depending on the application, a satisfactory frequency response over a specified bandwidth. However, the size (diameter) of the element will ultimately limit the bandwidth of its response. Because of this, characteristic disadvantages include increasingly (with frequency) narrow HF directivity patterns [8] and irregular HF response curves [24], along with low acoustic power output. Therefore, with the possible added nonexistence of point-source time-alignment, in a multi-planar system, parameters such as HF phase coherency and frequency response, along with directivity, may be irregular and unpredictable. Furthermore, these principal issues associated with multi-planar units, such as dodecahedrons, are directly related to the unpredictable HF behavior and instability within the individual direct-radiator element itself.

A primary issue of time alignment relates to the general, uniform motion of the entire diaphragm. The cone, or diaphragm, most commonly made of paper (but may also be made of plastic, metal, a synthetic material such as Kevlar, or a composite derivation), is sufficiently stiff at lower frequencies so that all of its parts move completely in phase as a single unit. Using the circular piston and infinite baffle model, the loudspeaker diaphragm may be considered, at lower frequencies, to be a vibrating piston of radius "a," moving with uniform velocity over its entire surface. This is a fair approximation at frequencies for which the axial distance "b," from the inside (apex) to the outside (base) of the cone, is less than about one-tenth of the signal wavelength [20] (see Figure 3.1). As an example, an axial distance of 3 cm will produce a uniform diaphragm velocity at frequencies up to approximately 1,143 Hz. Beyond this "critical" frequency, various sections of the diaphragm begin to move at different velocities. This is usually a complication with a larger diaphragm operating at a higher frequency. It has also been suggested that in units where the velocity in the diaphragm is greatly different from that in air, the differential component results in a mismatch at the diaphragm/air interface which may result in phase cancellation at certain frequencies [25]. This problem becomes more complex if sections of the diaphragm are moving at different velocities.

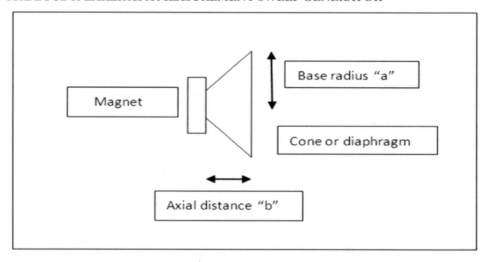

Figure 3.1: Profile of direct-radiator cone assembly.

A second issue with direct-radiator loudspeakers relates to the radiation pattern of the diaphragm, which is determined by the relationship between the base diameter "2a" (see Figure 3.1), and the emitted signal. Specifically, a vibrating piston mounted in an "infinite" baffle, whose diameter is less than one-third of the signal wavelength (ka < 1.0 with k = $2\pi/\lambda$, where λ is the signal wavelength), is essentially non-directional at low frequencies [12, 20].

Therefore, a typical subwoofer with a diameter of 45.7 cm (18 inches) is non-directional up to approximately 250 Hz. Keeping this in mind, it is interesting to note that, in active crossover sound reinforcement applications, the low-pass filter (LPF) crossover- frequency for a subwoofer diaphragm is usually anywhere between 80 and 120 Hz, depending on specific loudspeaker parameters, such as size. Consequently, due to this nondirectional characteristic of large diaphragms, several "identical" elements may be mounted on a planar cabinet baffle in order to improve the output efficiency at low frequencies. However, at higher frequencies, as the ratio between the loudspeaker diameter and signal wavelength increases (increasing frequency), there is also a corresponding increase in directivity, unpredictability, and therefore, phase interference between the diaphragms.

Finally, a third issue regarding the direct-radiator class of loudspeakers, and perhaps the most complex, concerns specific modes of resonance or deformation that force the loudspeaker diaphragm to break up into different regions of non-uniform motion and phase. These characteristic modes of resonance or destructive interference depend on parameters such as the diaphragm material and size (stiffness-to-mass ratio). Therefore, these modes will also vary from loudspeakers of one manufacturer to another, but in general, begin to occur in the approximate range of 300 to 1,000 Hz [20]. The result is a departure from uniform piston-like operation as the different areas of the diaphragm, separated by nodal lines or areas, start vibrating out of phase with the main drive or excursion and with one another (see Figure 3.2). This produces associated irregularities in amplitude, phase, and directivity. As frequency is increased, the loudspeaker breaks up into still different characteristic modes of vibration, producing different sets of nodal lines. Adding to the complexity is the fact that these changes occur with great rapidity as a function of frequency.

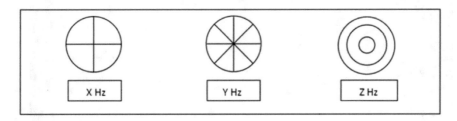

Figure 3.2: Nodal lines representing characteristic modal patterns of vibration, for a single-cone diaphragm, with neighboring sections having opposite phase.

The effects described above result in irregularities at higher frequencies and essentially produce relatively different amounts of sound radiated in different directions. The resulting sound field is diffuse, generating an unpredictable and inconsistent HF beam-width pattern. These issues of phase, amplitude, directivity, and frequency response are largely why the mid-to-high frequency response of a single, and therefore, a group of direct-radiator loudspeakers, is unstable and difficult to predict or measure accurately.

An HF tweeter element is of the direct-radiator category. Many near-field studio monitors incorporate dome tweeters in their design. Domes are the most common form of HF diaphragms, and are available in a wide variety of shape, size, and material. For HF reproduction, they are typically made small in order to reproduce higher frequencies with reasonable accuracy. However, the dome tweeter is a direct-radiator type of loudspeaker. Therefore, even though it may possess good sound quality, its efficiency, acoustic power reproduction, and directional control are limited. In a recording studio environment, these issues are not of much concern, as the sound is not required to travel a long distance. In a live or sound reinforcement environment where increased acoustic power efficiency and constant coverage are desired, a compression driver and horn may be more suitable for HF reproduction.

It has previously been mentioned that a loudspeaker of the direct-radiator class becomes increasingly directional and unpredictable as the ratio of the speaker diameter to the signal wavelength increases. Therefore, these types of speakers typically suffer from great deficiencies and irregularities in coverage or pattern control in the mid- to high-frequency regions. As illustrated in Figure 3.3, the beam-width (−6 dB corners) plot for the example direct-radiator, HF tweeter-based, three-way, near-field system (JBL 4412 with HF dome radiator) rolls off considerably with increase in frequency. The JBL 4412, first available in the mid-1980s but recently discontinued, possesses a frequency response within ±2 dB from 45 Hz to 20 kHz, with an HF crossover point of 4.5 kHz [26].

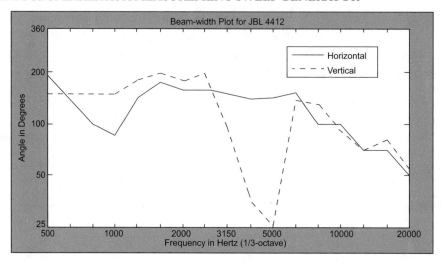

Figure 3.3: Angle of projection beam-width (−6 dB corners) vs. frequency plot for the JBL 4412, three-way, near-field, HF tweeter direct-radiator monitor system [26].

3.2.2 HORN LOUDSPEAKERS

A horn loudspeaker may effectively be described as an acoustic transformer—a device to increase the efficiency of sound radiation; that is, it functions as an acoustic impedance transformer of the air load at the throat of the horn and the corresponding air load at the mouth of the horn, the latter of which is directly coupled to the surrounding air mass. Horn loudspeakers take many forms depending on the required frequency response. For HF applications, the piston-like compression driver unit essentially consists of a small direct-radiator loudspeaker dome unit that is acoustically coupled or matched to the throat impedance of a flaring horn (see Figure 3.4). Almost any value of acoustic impedance may be produced through proper geometrical design of the horn's throat; thus, the match to the driver's diaphragm may be optimized [25]. This increased efficiency of the compression driver and horn design is a distinct advantage over the direct coupling of a cone or dome direct-radiator loudspeaker with the immediate surrounding air mass.

Compared to the low efficiency of a direct-radiator loudspeaker, the efficiency of radiation from a horn loudspeaker varies between 10% and 50% [20]. One of the primary functions of a horn is to deliver a large amount of acoustic power in a controlled, directional manner, with suitably low levels of nonlinear distortion. Therefore, a horn-loaded loudspeaker is designed to project sound over longer distances than the direct-radiator type of loudspeaker. When recording a loudspeaker-driven sine-sweep in a far-field environment such as a large concert hall, this may be a desirable characteristic in order to obtain characteristic orders of reflections along with a correspondingly more consistent and predictable wave-front.

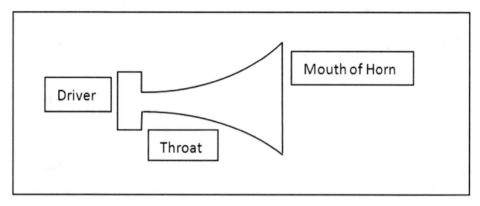

Figure 3.4: Profile of compression driver and horn loudspeaker. Reproduced, with kind permission from IEEE, from D. Frey, V. Coelho, and R.M. Rangayyan, "The loudspeaker as a measurement sweep generator for the derivation of the acoustical impulse response of a concert hall," *IEEE Canadian Conference on Electrical and Computer Engineering (CCECE),* Niagara Falls, ON, Canada, 4–7 May, 2008, pp. 301–304. © IEEE.

Aside from robustness and greater efficiency, a compression driver and CD horn combination can provide a much more uniform and predictable pattern of coverage than a direct-radiator loudspeaker. Once the coverage pattern is computed at a given frequency, the same coverage applies at all frequencies within the specified range. The beam-width plot for an example HF horn-based studio monitoring system is shown in Figure 3.5 (JBL 4430 with a bi-radial 100° × 100° horn). The coverage above the 1,000 Hz HF crossover frequency is reasonably uniform and consistent. Note that the −3 dB crossover point for a single HF horn, in a multi-way system, is commonly set between 1,000 and 2,000 Hz. The JBL 4430 two-way system, first made in 1981 and discontinued in 1999, was originally intended to be a horn-based alternative for accurate studio monitoring, providing CD pattern control and robustness. The JBL 4430 was designed for use in large monitoring environments and provides a frequency response within ±3 dB from 35 Hz to 16 kHz [27].

The Phasing Plug

An important section of the HF compression driver unit is the phasing plug which acts as part of the front chamber interface between the small direct-radiator driver diaphragm and the matching throat of the horn. At high frequencies, the wavelength of the sound is small in comparison to the diameter of the diaphragm. Large values of the ratio between the loudspeaker diameter and signal wavelength cause phase distortion or time-alignment problems within different sections of the diaphragm. Consequently, for any given frequency, there will be path-length variations resulting in different times of arrival at the central on-axis point in front of the diaphragm leading to the throat of the horn. Furthermore, a larger diaphragm will produce larger amounts of path-length variations.

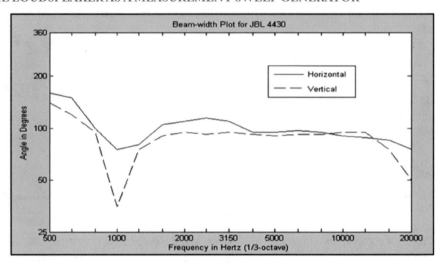

Figure 3.5: Angle of projection beam-width (−6 dB corners) vs. frequency plot for the JBL 4430, two-way, HF bi-radial horn loudspeaker system [27].

In addressing this problem, the phasing plug is used to alter the diaphragm-to-horn path length of waves emanating from different areas of the speaker element so that their phase is coherent [12]. Therefore, a phasing plug helps to minimize phase cancellations that would otherwise occur before the sound enters the throat of the horn. This contributes to the higher levels of efficiency that are characteristic of compression driver/horn loudspeakers.

3.3   HARMONIC DISTORTION IN LOUDSPEAKERS

Both direct-radiator and horn loudspeaker classifications produce their own characteristic variations of nonlinear harmonic distortion. As an example, a compression driver and horn loudspeaker will typically produce a higher level of second harmonic distortion [24]. This characteristic second-harmonic effect is essentially a consequence of high levels of sound or SPLs being reproduced by the compression driver element (diaphragm) coupled to the matching high compression ratio in the narrow throat of the flaring horn. This source of distortion is of particular concern in HF driver and horn design since the need for greater efficiency at high frequencies usually results in the use of higher compression ratios [12].

To further illustrate the effects of harmonic distortion, Figures 3.6 and 3.7 present experimental AIRF deconvolution measurement results between the recorded 15-second ESS measurement sweep and the corresponding ESS inverse filter, using the "convolve with clipboard" module developed by Farina [16]. The figures represent two different types of measurement loudspeakers: the Meyer MTS-4 HF horn loudspeaker system and the Dynaudioacoustics AIR 15 HF dome tweeter near-field monitor system. Both measurements were performed in the Eckhardt-Gramatté Hall at an on-axis distance of 6.8 m (row five) from the respective measurement loudspeaker. In both figures, the primary response is the desired linear impulse response of the hall. Preceding the primary response are the increasing orders—from right to left—of harmonic distortion, each represented by its own impulse response [8, 15]. The experiment was designed to drive both types of loudspeaker systems to the upper levels of their operating range, in order to obtain optimal and robust test signals to excite the hall. Of interest is the level of the second harmonic distortion component with the Meyer MTS-4 horn loudspeaker of Figure 3.6; as mentioned, this is primarily due to the high SPL through the compression driver and horn. High pressures, if propagated through a horn with little reduction in level, will generate high amounts of second harmonic distortion [28].

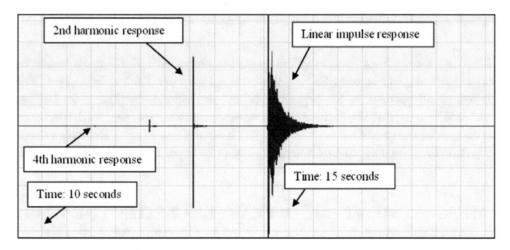

Figure 3.6: AIRF measurement result using the Meyer MTS-4 HF horn loudspeaker system as the 15-second ESS measurement signal generator. Time-scale: 0.5 s per division; amplitude-scale: 3 dB per division. Reproduced, with kind permission from IEEE, from D. Frey, V. Coelho, and R.M. Rangayyan, "The loudspeaker as a measurement sweep generator for the derivation of the acoustical impulse response of a concert hall," *IEEE Canadian Conference on Electrical and Computer Engineering (CCECE)*, Niagara Falls, ON, Canada, 4–7 May, 2008, pp. 301–304. © IEEE.

Representations according to Figures 3.6 and 3.7 present interesting nonlinear behavior of each loudspeaker type or class, while allowing for easy separation and documentation of the primary linear response. Therefore, because they can be isolated from the linear impulse response, nonlinear harmonic responses of any loudspeaker type or classification are not necessarily of primary concern, even though they provide interesting information for further analysis and documentation. This is one important advantage of the ESS measurement method.

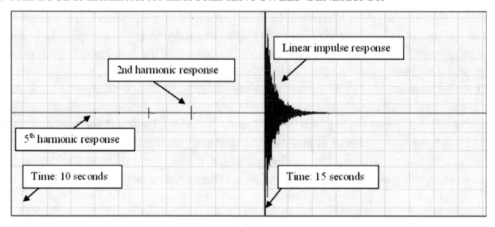

Figure 3.7: AIRF measurement result using the Dynaudioacoustics AIR 15 HF dome tweeter near-field monitor system as the 15-second ESS measurement signal generator. Time-scale: 0.5 s per division; amplitude-scale: 3 dB per division. Reproduced, with kind permission from IEEE, from D. Frey, V. Coelho, and R.M. Rangayyan, "The loudspeaker as a measurement sweep generator for the derivation of the acoustical impulse response of a concert hall," *IEEE Canadian Conference on Electrical and Computer Engineering (CCECE)*, Niagara Falls, ON, Canada, 4–7 May, 2008, pp. 301–304. © IEEE.

3.4   MEASUREMENT LOUDSPEAKERS USED IN THE WORK

3.4.1   THE MEYER MTS-4 HF HORN LOUDSPEAKER SYSTEM

The Meyer MTS-4 loudspeaker system [23] is used in the present work as one of two experimental sources to generate a sine-sweep source measurement signal in order to measure an AIRF. The Meyer MTS-4 has a 1,000 Hz crossover point separating the MS-12 12-inch direct-radiator cone and the MS-2001A compression driver. The subsequent HF horn possesses a coverage or CD pattern of 70 degrees horizontal by 60 degrees vertical. This fully integrated and modular four-way loudspeaker system is self-powered and DSP time-aligned with an amplifier power rating of 620 W/channel. In addition, its frequency response is within ±3 dB from 30 Hz to 16 kHz [23]. Even with its extended LF response, it is evident that the HF response is limited—a typical horn design compromise for an increase in acoustic power output while maintaining a smooth frequency response. Independent amplifier and control electronics are provided in the cabinet for three LF direct-radiator-type cone speakers (18", 15", and 12"), and the corresponding HF compression driver and horn (4" driver diaphragm and 2" horn throat).

3.4.2   THE DYNAUDIOACOUSTICS AIR 15 HF TWEETER NEAR-FIELD MONITOR

The Dynaudioacoustics AIR 15 is the second experimental source used in the present work. In contrast to the Meyer MTS-4 system, the AIR 15 is a professional, near-field, studio and/or mastering monitor loudspeaker. It is a self-powered, DSP-controlled, direct-radiator-type loudspeaker system, incorporating a 254 mm (10") woofer and a 28 mm (1.1") HF soft-dome tweeter. This high-quality loudspeaker possesses a frequency response within ±3 dB from 33 Hz to 22 kHz [29], and provides amplifier power of 200 W/channel for its two-way operation.

3.5   PERFORMANCE COMPARISON OF THE MEASUREMENT LOUDSPEAKERS

For comparison purposes, each loudspeaker used in the project was centered near the front edge of the stage in the Eckhardt-Gramatté Hall. At row five (6.8 m from the stage), the level of the Meyer MTS-4 HF horn system (for the 22 Hz–22 kHz ESS) was measured to be a continuous average of 100 dB SPL. Therefore, at 1.7 m, the loudspeaker is operating at a continuous average of approximately 112 dB (6 dB per doubling of distance for a point source using the inverse square law). According to specifications, the rated maximum SPL of the Meyer MTS-4 is a peak 140 dB at 1 m. In order to design a loudspeaker pre- or inverse filter, pink noise spectral analysis of the MTS-4 was performed in a separate session, with the measurement being performed at similar distances and levels of 1.7 m and 112 dB SPL. The intention was to maintain a consistent SPL between the spectral analysis and the later ESS recording in the hall, where the prefilter was inserted and used to maintain a flat amplitude response of the MTS-4 loudspeaker output. It was determined that the amplitude response of the AIR 15 was adequately flat and did not need a separate pre-filter.

A single Dynaudioacoustics AIR 15 HF tweeter near-field monitor measured a continuous average of 88 dB SPL at row five. Therefore, at 1.7 m, the loudspeaker is operating at a continuous average of approximately 100 dB SPL. Technical specifications for the AIR 15 indicate a maximum SPL at 1 m (IEC Long Term) of 103 dB (Root-Mean-Squared [RMS] value). As mentioned previously, the experiment was designed to drive both types of loudspeaker systems to the upper levels of their operating range, in order to obtain optimal and robust test signals to excite the hall.

3.5.1   FREQUENCY RESPONSE

Normalized (−1 dB) and edited mono recordings for each project loudspeaker, performed at row five (6.8 m) of the Eckhardt-Gramatté Hall, are illustrated in Figures 3.8 and 3.9. Both recordings exhibit peaks in the extreme low and lower midrange frequencies of the 15-second, 22 Hz–22 kHz ESS. It is interesting to note that, through prior spectral analysis of the hall, these frequency ranges were observed to be acoustically excessive. The recordings are in agreement with this observation.

From Figure 3.9, it also appears that the response of the AIR 15 HF tweeter near-field monitor is smoother or less variable in amplitude at the upper-midrange and higher frequencies. This is partly

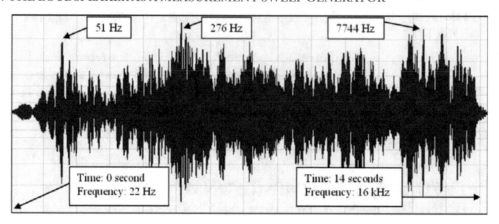

Figure 3.8: Normalized (−1 dB) and edited 15-second ESS recording performed at row five of the Eckhardt-Gramatté Hall using the Meyer MTS-4 HF horn loudspeaker system. Time-scale: 1 s per division; amplitude-scale: 3 dB per division. Reproduced, with kind permission from IEEE, from D. Frey, V. Coelho, and R.M. Rangayyan, "The loudspeaker as a measurement sweep generator for the derivation of the acoustical impulse response of a concert hall," *IEEE Canadian Conference on Electrical and Computer Engineering (CCECE)*, Niagara Falls, ON, Canada, 4–7 May, 2008, pp. 301–304. © IEEE.

due to the extended HF response of the AIR 15 (22 kHz) dome tweeter compared to the HF −3 dB roll-off frequency of 16 kHz with the MTS-4 horn. As previously mentioned, a horn loudspeaker commonly trades off an extended HF response for an increase in acoustic power output. In addition, a common and traditional criticism of HF horn loudspeakers has been that their frequency response is not as smooth (flat) as that of a high-quality HF tweeter element [24]. This is due to the traditional horn design compromise that maintaining a constant pattern of coverage, independent of frequency, is accomplished at the expense of a flat or constant on-axis amplitude response with frequency.

3.5.2   EFFICIENCY, DIRECTIVITY, AND MEASUREMENT SNR

Acoustic measurement in a concert hall is primarily a far-field application. Therefore, a near-field monitor is not the proper choice as a measurement loudspeaker in this type of environment. Aside from the characteristic inefficiency and lower acoustic power of direct-radiator elements, the diffusive radiation characteristics of an HF tweeter near-field monitor prevent the sound from being projected to the more remote locations of the hall. Consequently, it was observed that, even at Row 10 of the Eckhardt-Gramatté Hall, the on-axis SNR of the recording with the AIR 15 begins to deteriorate as a higher recording level is needed to compensate for its short-throw pattern. However, this is to be expected. The characteristic wider diffusion pattern of an HF tweeter-based near-field monitor is appropriate in the smaller environment of a recording studio where distance is not an issue.

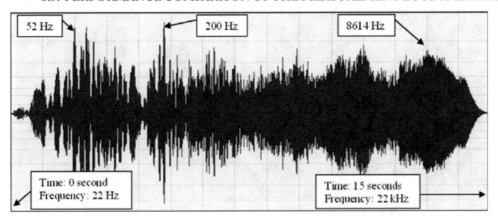

Figure 3.9: Normalized (−1 dB) and edited 15-second ESS recording performed at row five of the Eckhardt-Gramatté Hall using the Dynaudioacoustics AIR 15 HF dome tweeter near-field monitor system. Time-scale: 1 s per division; amplitude-scale: 3 dB per division. Reproduced, with kind permission from IEEE, from D. Frey, V. Coelho, and R.M. Rangayyan, "The loudspeaker as a measurement sweep generator for the derivation of the acoustical impulse response of a concert hall," *IEEE Canadian Conference on Electrical and Computer Engineering (CCECE)*, Niagara Falls, ON, Canada, 4–7 May, 2008, pp. 301–304. © IEEE.

In contrast, through its higher levels of efficiency, output power, and constant directional coverage, the HF horn-loaded MTS-4 maintained a high SNR at row ten, continuing to provide an excellent recording for documentation and processing, without the need for any additional synchronous averaging or noise reduction software and the associated issues of processing integrity. Figures 3.10 and 3.11 provide graphic representations of the respective recorded system noise levels between successive ESS sweeps for each type of loudspeaker-generated recording performed at Row 10.

In maintaining equal, normalized, digital recording levels of −1 dB for each loudspeaker-generated sweep, measurements from Figure 3.10 indicate that the amplitude of the recorded system noise for the MTS-4 reached a peak level of −42 dB and a respective average level of −53 dB (average RMS power), whereas measurements of the AIR 15 from Figure 3.11 indicate a respective peak level of −29 dB and an average level of −40 dB (average RMS power). For purposes of comparison, similar noise level measurements were taken at row five. At row five, the MTS-4 reached a peak level of −37 dB (−50 dB average RMS power), whereas the AIR 15 measured a peak amplitude of −34 dB (−45 dB average RMS power). It is interesting to note that the sweep recording generated by the MTS-4 maintained and even improved on the system noise measurement at row ten, whereas the recording generated by the AIR 15 suffered a noise increase of 5 dB, in both peak and average RMS power.

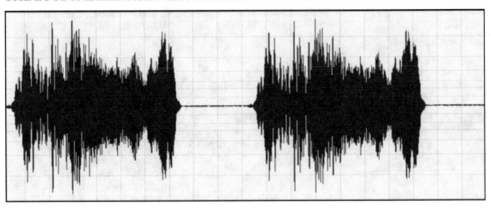

Figure 3.10: Normalized (−1 dB) Meyer MTS-4 HF horn-generated 15-second ESS "train" recording performed at row ten of the Eckhardt-Gramatté Hall. Time-scale: 2 s per division; amplitude-scale: 3 dB per division. Magnitude of system noise is illustrated between the two sweeps.

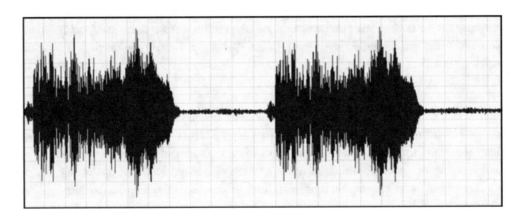

Figure 3.11: Normalized (−1 dB) Dynaudioacoustics AIR 15 HF tweeter-generated 15-second ESS "train" recording performed at row ten of the Eckhardt-Gramatté Hall. Time-scale: 2 s per division; amplitude-scale: 3 dB per division. Magnitude of system noise is illustrated between the two sweeps.

Similar recorded system noise measurements were performed at other rows, including the rear, of the Eckhardt-Gramatté Hall. The results are listed in Table 3.1 and plotted in Figure 3.12. Aside from Row 5, there is an average difference of about 10 dB per row between the two loudspeakers. At Row 10, the difference is 13 dB, as previously mentioned.

Table 3.1: Average RMS power (in dB) of the recorded system noise in the Eckhardt-Gramatté Hall

Row no.	5	7.5	10	12.5	15	rear
Noise (dB)						
AIR 15	-45	-45	-40	-44	-43	-40
MTS-4	-50	-54	-53	-53	-55	-51

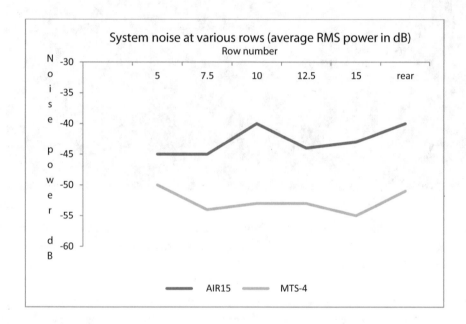

Figure 3.12: Recorded system noise at various rows in the Eckhardt-Gramatté Hall for each project loudspeaker. System noise amplitude is average RMS power in dB.

3.6   ANECHOIC SIMULATIONS OF THE MTS-4 MEASUREMENT LOUDSPEAKER

In order to provide a more comprehensive AIRF measurement over 360 degrees horizontally, it would be possible to rotate the Meyer MTS-4 loudspeaker and record on-axis sweeps at 45-degree intervals. In this manner, there would be a document and resulting AIRF for each location. Furthermore, a single AIRF of the hall could be determined through synchronous averaging. Figure 3.13 presents the horizontal radiation pattern of a single Meyer MTS-4 at a one-octave relative band-width centered at 1 kHz. It is evident that the loudspeaker performs well in a far-field environment. This virtual anechoic simulation was performed using the Meyer Sound Multipurpose Acoustical Prediction Program (MAPP) Online Pro measurement software [30].

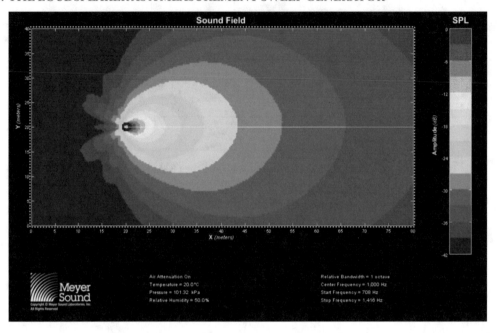

Figure 3.13: Horizontal radiation pattern of a single Meyer MTS-4 at a one-octave relative bandwidth centered at 1 kHz. Length of plot (x): 5 m per division (80 m); width of plot (y): 5 m per division (40 m). SPL colors are in 3 dB decrements.

Using the same 1 kHz reference band, Figure 3.14 presents the horizontal pattern of eight MTS-4 loudspeaker positions each rotated at 45 degrees. The pattern is not perfectly spherical but it does represent the symmetry and consistency possible with multiple horn-loaded loudspeaker positions in a far-field environment. This creates the ability to produce a wider radiation pattern if necessary. The MAPP software enables a loudspeaker arrangement to be tested and performed virtually on a computer beforehand.

Figure 3.15 illustrates the wide frequency response of the MTS-4 from 16 Hz to 16 kHz with the virtual measurement microphone positioned on axis at 2 m from the loudspeaker. Meyer Sound offers the MAPP Online Pro virtual measurement software [30] as part of their continually developing library of Meyer loudspeaker measurements performed originally in their own anechoic chambers and subsequently offered as virtual simulations in the software.

3.7   REMARKS

Compression driver and horn loudspeaker systems provide improvements in efficiency (transfer of electrical energy to acoustical energy) and radiation or directional control over their direct-radiator counterparts. Furthermore, due to the phasing plug, they do not suffer to the same degree from the

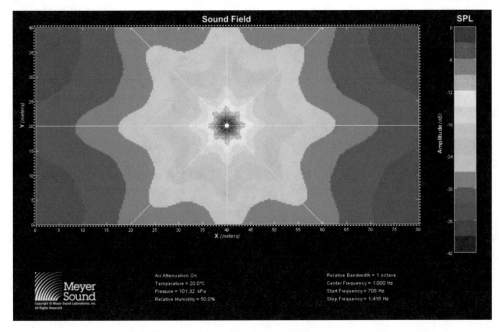

Figure 3.14: Horizontal radiation pattern of eight MTS-4 loudspeaker positions each rotated at 45 degrees. Length of plot (x): 5 m per division (80 m); width of plot (y): 5 m per division (40 m). SPL colors are in 3 dB decrements.

problems of diaphragm output phase distortion or time-domain coherency produced by the physical and modal deformations associated with direct-radiator loudspeakers. These are important factors if the measurement source loudspeaker is to be as transparent as possible; that is, the frequency response of the output of the loudspeaker is equal to the frequency response of the input. This consistency helps to ensure that the negative effects of the measurement loudspeaker on the AIRF are greatly reduced.

The present work has shown that an HF horn-based loudspeaker system provides some advantages as a measurement sweep generator for the derivation of the AIRF of a music performance hall. One of these advantages is the ability to radiate the measurement sweep to the more remote locations of a hall. This provides the ability to stimulate and capture higher orders of reflections that are a critical part of the identity of the hall [22]. A similar example would be the proper stimulation and capture of the upper harmonics of a musical instrument in order to document the characteristics or identity of that instrument.

Consequently, even at the rear of the 384-seat Eckhardt-Gramatté Chamber Music Hall at the University of Calgary, the horn-loaded Meyer MTS-4 consistently excites the hall with a robust and articulate ESS measurement signal (see Figure 3.13). The corresponding on-axis measurements of

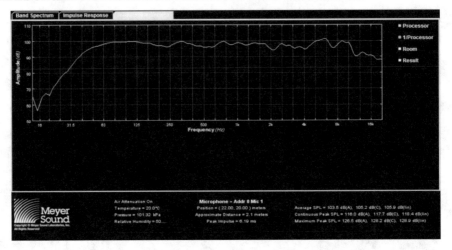

Figure 3.15: Frequency response of the Meyer MTS-4 using MAPP [30]. Frequency-scale: 16 Hz to 16 kHz (logarithmic octaves); amplitude-scale: 10 dB per division (50 to 110 dB).

recorded system noise at various rows (see Table 3.1 and Figure 3.12) provide numerical confirmation that the improvements in efficiency and sound radiation of a compression driver and horn loudspeaker system can provide an unwaveringly higher SPL at the greater distances encountered in a far-field environment.

However, the weakest link in the acoustic characterization or measurement process is still the reference loudspeaker. Transparency is the ultimate goal to ensure that the loudspeaker will not influence the measurement of an AIRF. Horn (or waveguide) loudspeakers and their far-field capabilities offer some advantages over direct-radiator loudspeakers and their near-field limitations. With continuing improvements in DSP control, and corresponding advances in compression driver and horn technology, it is now possible to realize some of the potential benefits in acoustical measurement associated with the increased definition, control, predictability, and output capability of a loudspeaker horn system.

CHAPTER 4

Convolution and Filtering

DSP has led to both increased efficiency and increased difficulty in measuring the AIRF of a music performance hall. The efficiency of DSP produces an environment of increased detailed analysis but also produces greater difficulty in achieving accurate numerical quantification. An initial difficulty is the necessary conversion of an analog signal in continuous time to that of a digitally sampled representation in discrete time. This is called analog-to-digital (A/D) conversion. Convolution and filtering are also important in signal processing and the derivation of an AIRF. In order to achieve an accurate numerical AIRF representation, it is necessary to understand the numerical processing involved in the execution of a particular convolution or filtering algorithm. This is first accomplished through a deeper understanding of signal processing basics, such as the discrete-time or digital representation of continuous-time signals. Other basics include the advantages of the Fourier transform and processing in the frequency domain as compared to processing in the time domain. It is then important to look at some of the different traditional convolution methods, such as overlap-save, and to develop an understanding of some of the corresponding convolution algorithms and techniques available for both implementation and optimization.

The process of deconvolution and its associated difficulties have been the subjects of a vast literature spanning the numerous special fields of science [31]. This is because there are numerous methods of deconvolution available that are employed in various disciplines such as telecommunications, biomedical image analysis, and acoustic measurement. The difficulty with deconvolution is that it is essentially a problem of deconstruction. That is, it involves the separation of a signal or signals from a composite filtered result, which often includes independently generated additive noise. Therefore, the efficiency of a particular method is often dependent on information previously available about the signals involved. Consequently, there are processing constraints imposed in order to increase the efficiency and provide methods for optimization. Constraints become even more important when there is a substantial amount of noise present. If not determined carefully, such constraints will also negatively affect the accuracy of the result. However, before we discuss deconvolution and its associated difficulties, it is important to start with some basics of signal processing.

4.1 DISCRETE-TIME PROCESSING OF CONTINUOUS-TIME SIGNALS

For a Linear Time-Invariant (LTI) system, a representation of the input sequence as a weighted sum of delayed impulses leads to a representation of the output as the aggregate sum of delayed impulse responses [32]. Whether the system is electrical, mechanical, or acoustical, convolution is performed between the input signal, $x(t)$, and the system's characteristic impulse response function, $h(t)$, producing an output signal, $y(t)$. This makes convolution one of the most important and traditional operations in signal processing (see Figure 4.1).

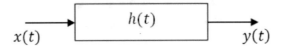

$$x(t) \qquad h(t) \qquad y(t)$$

Figure 4.1: Convolution in an LTI system.

An illustration of natural, everyday acoustical convolution is attending a concert for solo violin, and hearing the continuous-time convolution of the system LTI impulse response (concert hall acoustics) and the input signal (violin). The resulting output, $y(t)$, is simply the reverberation or acoustical impulse response of the hall, $h(t)$, acting on the signal produced by the musical instrument, $x(t)$. It must be emphasized that in this example, the signals are analog (continuous-time) and so, theoretically, the convolution integral

$$y(t) = x(t) * h(t) = \int_{-\infty}^{\infty} x(\tau)h(t - \tau)d\tau \qquad (4.1)$$

is used to represent the system in a mathematical manner. Generally, the asterisk denotes linear or aperiodic convolution as opposed to circular or periodic convolution, which will be discussed shortly.

Even though the integral in Equation (4.1) presents a useful mathematical concept, it can be difficult to implement numerically, as it is a function of the area under the curve. The implementation becomes numerically practical by sampling the continuous-time signal and converting it to a discrete-time signal, where the signal amplitudes are also discrete and represented as a sequence of numbers or quantization levels. In the digital representation of a signal, both time and amplitude are discrete [32]. A major application of discrete-time systems is in the processing of continuous-time signals. This is accomplished through the implementation of a system with the general form as shown in Figure 4.2. Consequently, for computer numerical analysis, the convolution integral is replaced by the more practical discrete-time convolution sum

$$y[n] = x[n] * h[n] = \sum_{k=-\infty}^{\infty} x[k]h[n - k]. \qquad (4.2)$$

Even though it is common to process an analog signal digitally, there are many problems with considering a digital signal to be an accurate representation of its analog counterpart. One difficulty is that, in digital sampling, the height or amplitude of each impulse needs to represent a particular area of the continuous-time or analog waveform. Quantization error is another accuracy issue with discrete-level representation of an analog variable in discrete time. Therefore, the convolution sum in Equation (4.2) should be considered only as a discrete-time representation of the integral in

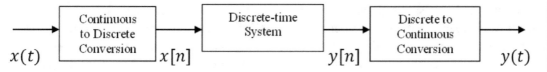

Figure 4.2: Discrete-time processing of continuous-time signals.

Equation (4.1). The convolution integral plays mainly a theoretical role in continuous-time linear system theory; the convolution sum, in addition to its theoretical importance, serves as an explicit realization of a discrete-time linear system [32]. Ultimately, even though it is not perfect, digital sampling in discrete time is a convenient method for numerical analysis in the computer domain.

4.1.1   FINITE-LENGTH SEQUENCES

In Equations (4.1) and (4.2), there are no finite limits on the intervals of integration and summation. One of the properties of stability for any LTI discrete-time system is that its output must be absolutely summable [32]; that is, it must be a convergent series or have a finite-length interval. With N representing a finite number of data points, consider a discrete-time LTI system with the impulse response

$$h[n] = u[n] - u[n - N] = \begin{cases} 1, & 0 \le n \le N - 1 \\ 0, & \text{otherwise,} \end{cases}$$

and with the input

$$x[n] = a^n u[n], \qquad 0 < a \le 1,$$

where $u[n]$ is the unit step function

$$u[n] = \begin{cases} 1, & n \ge 0 \\ 0, & \text{otherwise.} \end{cases}$$

Then, the result of convolution is finite and absolutely summable:

$$y[n] = x[n] * h[n] = \sum_{k=n-N+1}^{n} a^k u[k] \qquad \text{for} \quad n \ge 0. \tag{4.3}$$

4.1.2   FREQUENCY-DOMAIN REPRESENTATION

In signal processing, it is often necessary to convert time-domain signals into transformed representations that are more convenient to analyze and process. In addition, the respective theorems and properties of such transforms provide greater understanding of the nature of different signals and systems. The complex s-domain and z-domain representations of the Laplace transform and z-transform are examples of such transforms [32]. Another interesting transformation is the frequency-domain (or spectral) representation via the continuous Fourier transform (FT) and its discrete counterpart, the discrete Fourier transform (DFT). Fourier analysis is essential for describing certain types of systems and their properties in the frequency domain. For convenience and greater insight, there are also a variety of theorems that relate operations on a time-domain sequence to operations on the respective DFT. The DFT is a discrete, or digital representation, not an approximation, of the continuous FT. DFT representations of time-domain signals can lead to a dramatic reduction in the computational requirements for specific operations, such as convolution. The convolution theorem of the FT essentially states that the process of convolution in the time domain is equal to a simple multiplication in the frequency domain, that is

$$x(t) * h(t) \longleftrightarrow X(e^{j\omega})H(e^{j\omega}), \qquad \omega = 2\pi f, \tag{4.4}$$

where $X(e^{j\omega})$ is the FT of $x(t)$ and f is frequency in Hz.

The corresponding convolution theorem for the DFT is

$$x[n] * h[n] \longleftrightarrow X(e^{j\Omega})H(e^{j\Omega}), \qquad \Omega = \frac{2\pi k}{NT}, \tag{4.5}$$

where T is the sampling interval and k is the frequency sample index, with $k = 1, 2, \ldots, N$.

The analog notation in Equation (4.4) refers to continuous-time processing with an unlimited duration, as in the convolution integral in Equation (4.1). On the other hand, Equation (4.5) refers to discrete-time processing with a finite-length sequence, or a discrete number of samples, N, as in the finite-length convolution sum represented in Equation (4.2). The convolution theorem illustrates the complex exponential representation of an LTI system using the Fourier transform—an important relationship because complex exponential sequences are eigenfunctions of LTI systems, and therefore, may be used to represent them. This is a fundamental property of LTI systems [32].

4.1.3   COMPUTATIONAL REQUIREMENTS OF CONVOLUTION

In a numerical evaluation of the computational resources required for performing convolution in the frequency domain, it may be useful to first establish the number of multiplications required to perform the linear convolution of two, discrete, finite-length sequences in the time domain, using the convolution sum of products in Equation (4.2). Typically, for two N-point complex data sequences, the number of real multiplications required is $O[N^2]$, which is a function that represents the "order of" constant of proportionality [32]. Therefore, if N is large, the computational demands of this "brute force" convolution method can be considerable and even excessive. For example, let us consider a 15-second ESS measurement signal and its corresponding 15-second convolving ESS inverse filter. With a sampling frequency of 96 kHz, the excitation sweep and inverse filter each consist of $N = 1,440,000(96,000 \times 15)$ sampled data points. Therefore, for the two N-point complex data sequences, and even if, hypothetically, $O[N^2] = 1N^2$, the number of real multiplications required for linear convolution in time is still a considerable $N^2 = 2.0736 \times 10^{12}$. Implementing this type of an intensive procedure can be challenging, especially if it needs to be done on line with minimal latency and execution time. However, by using the time-frequency relationship expressed in Equation (4.5), it is possible to use efficient algorithms for computing the DFT of a sequence in order to make the computation of the results of convolution more practical. In particular, time-domain signals are transformed to the frequency domain through what are known collectively as fast Fourier transform (FFT) algorithms [32].

The DFT and its associated properties, such as Equation (4.5), play an important role in the analysis, design, and implementation of discrete-time signal-processing algorithms and systems. The FFT is simply an efficient algorithm for evaluating a DFT. By avoiding a large number of duplicate multiplications, the FFT is computationally much more efficient than a standard DFT [31] and produces an identical result. Indeed, the increased efficiency of the FFT reduces the number of real multiplications required to compute the DFT of an N-point complex signal from $O[N^2]$ to $O[N\log_2 N]$ [32]. Consequently, if N is large, such as the $N = 1,440,000$ data points for the 15-second ESS, the number of real multiplications required for a DFT is reduced significantly from 2.0735×10^{12} to approximately 2.9459×10^7. The "order of" constant of proportionality, $O[N^2]$, essentially depends on the type of DFT evaluation that is being implemented.

Therefore, the FFT offers a practical way of implementing an LTI system as it can also reduce the computational requirements of $O[N^2]$ needed for the discrete-time, linear convolution of Equation (4.2). However, it is important to note that the use of the FFT in the frequency domain to achieve convolution involves periodic or circular convolution (discussed in the next section). This is in contrast to the linear convolution that is desired in the time-domain representation of an LTI system. Furthermore, the periodic convolution of two signals does not generally equal the aperiodic or linear convolution of the same two signals [32, 33]. This is because, in periodic convolution, portions of the data sequences overlap in a manner different from that with linear convolution. To avoid this problem of wrap-around error, or time aliasing, and to make periodic or circular convolution achieved with the FFT produce the same result as linear convolution in time, the original sequences are simply padded with zeros. This is called zero-padding and is presented in the following section. The efficiency of FFT algorithms is so high, in fact, that in many cases, the most efficient procedure for implementing a convolution is to compute the transform of the sequences to be convolved, multiply their transforms, and then compute the inverse transform of the product of transforms [32]. The FFT conversion and convolution of two discrete-time sequences, according to Equation (4.5), is illustrated in Figure 4.3. It essentially involves: two forward FFT conversions, one N-point complex multiplication, and one inverse FFT (IFFT) conversion.

4.1.4 CIRCULAR CONVOLUTION VERSUS LINEAR CONVOLUTION

Suppose there are two discrete sequences as follows:

$$x[n] = \{1, 2, 3, 4\} \quad \text{and} \quad h[n] = \{4, 3, 2, 1\}$$
$$\text{Length } L = 4 \qquad\qquad \text{Length } P = 4$$

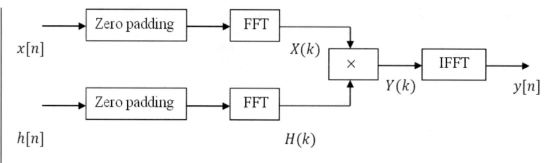

Figure 4.3: Convolution of two sequences using the FFT.

As presented earlier, linear time-domain convolution of two sequences involves $O[N^2]$, or in this example, $O[L \times P] = 16$ real multiplications and results in $L + P - 1 = 7$ samples [32]. One step of the computation for $n = 3$ is shown below.

$$
\begin{array}{ll}
1\ 2\ 3\ 4 & x[k] \\
1\ 2\ 3\ 4 & h[n-k] \quad \text{for} \quad n = 3 \\
\hline
\end{array}
$$

Multiply corresponding samples
and add terms: $1 + 4 + 9 + 16 = 30$ therefore,

$$
y[n] = \sum_{k=0}^{L-1} x[k]h[n-k] = 30 \qquad \text{for} \quad n = 3.
$$

The final result is

$$
y[n] = \{4, 11, 20, 30, 20, 11, 4\} \quad 0 \le n \le L + P - 2 = 6
$$
$$
\uparrow n = 3
$$

However, in using the FFT for the computation of a DFT, the two sequences $x[n]$ and $h[n]$ are considered to be periodic, that is,

$$
x_p[n] = \{\dots 1, 2, 3, 4, 1, 2, 3, 4, \dots\} \quad \text{and} \quad h_p[n] = \{\dots 4, 3, 2, 1, 4, 3, 2, 1, \dots\}
$$

Therefore, we have,

One period $N = 4$

$$\ldots 1\ 2\ 3\ 4\ 1\ 2\ 3\ 4\ 1\ 2\ 3\ 4\ \ldots \quad x_p[k]$$

$$\ldots 4\ 1\ 2\ 3\ 4\ 1\ 2\ 3\ 4\ 1\ 2\ 3\ \ldots \quad h_p[n-k]$$

↑

Multiply and
add terms over $N = 4$: $4 + 2 + 6 + 12 = 24$ for $n = 0$

$$y_p[n] = x_p[n] \bullet h_p[n] = \sum_{k=0}^{N-1} x[k]h[n-k], \quad N = L = P = 4$$

$$y_p[n] = \{\ldots 24, 22, 24, 30, \ldots\}, \quad 0 \le n \le N - 1.$$
↑ $n = 0$

$$y_p[n] = x_p[n] \bullet h_p[n] = \sum_{k=0}^{N-1} x[k]h[n-k], \quad N = L = P = 4$$

$$y_p[n] = \{\ldots 24, 22, 24, 30, \ldots\}, \quad 0 \le n \le N - 1.$$
↑ $n = 0$

Note that periodic convolution is defined only for two signals having the same duration or number of samples (N); the result also has the same number of samples, N, in each period. This example illustrates why the periodic convolution of two signals does not generally equal the aperiodic convolution of the same two signals. However, by zero-padding the original sequences, the resulting circular convolution using the FFT can be made to equal the result produced by linear convolution in time.

Consider the same two discrete sequences as before:

$$x[n] = \{1, 2, 3, 4\} \quad \text{and} \quad h[n] = \{4, 3, 2, 1\}$$
$$\text{Length } L = 4 \qquad\qquad \text{Length } P = 4$$

Pad the signals with zeros such that they have at least $N = L + P - 1 = 7$ samples, as

$$x_p[n] = \{1, 2, 3, 4, 0, 0, 0\} \quad \text{and} \quad h_p[n] = \{4, 3, 2, 1, 0, 0, 0\} \quad \text{with} \quad N = 7$$

Therefore, we now have, for $n = 0$, the following situation with the periodic versions of the zero-padded signals:

One period $\quad N = 7$

$$\ldots 4\ 0\ 0\ 0\ 1\ 2\ 3\ 4\ 0\ 0\ 0\ 1 \ldots \quad x_p[k]$$

$$\ldots 0\ 1\ 2\ 3\ 4\ 0\ 0\ 0\ 1\ 2\ 3\ 4 \ldots \quad h_p[n-k]$$

$$\uparrow$$

Multiply and
add terms over N = 7 : $\qquad\qquad 4\ 0\ 0\ 0\ 0\ 0\ 0 = 4 \quad$ for $\quad n = 0$

The resulting convolution is,

$$y_p[n] = x_p[n] \bullet h_p[n] = \sum_{k=0}^{N-1} x[k]h[n-k], \qquad N = 7$$

$$y_p[n] = \{4, 11, 20, 30, 20, 11, 4\}, \qquad 0 \leq n \leq N - 1.$$
$$\uparrow n = 0$$

It is evident that the result of periodic or circular convolution, $y_p[n]$, now equals the result of the linear convolution in time, $y[n]$.

The above example is, of course, a simplification. The original sequences $x[n]$ and $h[n]$ will usually not be of the same length, and therefore, will not be padded with zeros equally. This is because when two FFT sequences are multiplied together, they must contain the same number of coefficients, and therefore, the two zero-padded sequences must contain the same number of data points, $N \geq L + P - 1$.

In terms of multiplications and additions, FFT algorithms can be orders of magnitude more efficient than competing algorithms [11]. As an experiment to evaluate the speed of an FFT, MAT-LAB was used for both the aperiodic convolution, according to the "brute force" convolution sum, and periodic convolution, using the FFT, of two sequences $x[n]$ and $h[n]$, each containing 100,000 data points. The results were identical. However, the aperiodic convolution sum in the time domain was performed in 8.498 seconds. Comparatively, the periodic convolution using the FFT was performed in 0.573 second.

4.2   CONVOLUTION ALGORITHMS

Convolution is an effective way of implementing an LTI system (see Figure 4.1). There are a number of convolution algorithms available that provide efficient methods of obtaining a discrete-time linear convolution through the use of the FFT in the frequency domain. Procedures such as zero-padding permit the realization of linear convolution of two finite-length sequences using periodic or circular convolution implied by the use of the FFT. However, for signals that are very long in duration, using a single N-point FFT to filter a large number of data points may be impractical to compute. One solution to this problem is to use block convolution [32], in which the input signal to be filtered is divided into segments of, say, length L. Block convolution is essentially a modular procedure where each subsequent input section of length L is convolved separately with the finite-length impulse response of the filter being applied, of, say, length P. The shifted filtered sections are then fitted back together appropriately in a juxtaposition of output modules. In this manner, the linear filtering of each input block of length L can then be implemented using the FFT. One traditional block convolution procedure, that has been used in the convolution of audio signals, is the overlap-save method [32, 34].

4.2.1   THE OVERLAP-SAVE METHOD

The well-known overlap-save method of block convolution takes advantage of an important time-aliasing relationship produced by circular convolution. The entire input signal $x[n]$, of an unspecified duration or length, may be represented as a sum of shifted finite-length segments of length L; that is,

$$x[n] = \sum_{r=0}^{\infty} x_r[n - rL].$$

Specifically, for an L-point circular convolution, where the input block $x_r[n]$ is of length L, and the impulse response $h[n]$ of length $P(P < L)$, the first $(P - 1)$ points of the result $y_r[n]$ are corrupted by time-aliasing, while the remaining $(L - P + 1)$ points are not corrupted, and are what would have been produced using linear covolution.

To illustrate, suppose there are the two sequences, the input block $x_r[n]$ of length L, and the impulse response $h[n]$ of length P, as follows:

$$x_r[n] = \{1, 2, 3, 4\}, \quad L = 4 \quad \text{and} \quad h[n] = \{3, 2, 1\}, \quad P = 3.$$

It is important to note that, for an L-point circular convolution, $h[n]$ must be padded with zeros so that $P = L$. Therefore, $h[n] = \{3, 2, 1, 0\}$. The L-point result of circular convolution is

$$x_r[n] \bullet h[n] = y_r[n] = \{14, 12, 14, 20\}.$$

For comparison, performing the $(L + P - 1 = 4 + 3 - 1 = 6$ point) linear convolution of the same two sequences, without zero padding, produces

$$x_r[n] * h[n] = y_r[n] = \{3, 8, 14, 20, 11, 4\}.$$

In looking at the first $L = 4$ points only, we have

$$y_r[n] = \{3, 8, 14, 20\}, \quad n = 0, 1, 2, 3.$$

Due to the time-aliasing effect with circular convolution, the first $(P - 1 = 2)$ points of the L-point circular and linear results are not the same, but the remaining $(L - P + 1 = 2)$ points are. Therefore, a crucial characteristic of the overlap-save method is in performing an L-point circular convolution of an input block, and then extracting the first $(P - 1)$ points of the resulting output segment, $y_r[n]$, and concatenating the remaining $(L - P + 1)$ points of $y_r[n]$ to produce the complete output, $y[n]$.

The first $(P - 1)$ points of each input block, $x_r[n]$, of length, L will overlap the last $(P - 1)$ points of the previous input block. This ensures that the first $(P - 1)$ points being discarded from $y_r[n]$ have already been calculated in the previous output block and are not important. Consequently, for the input and output blocks we have,

$$x_r[n] = x[n + r(L - P + 1) - P + 1], \quad 0 \leq n \leq L - 1,$$
$$y_r[n] = y_r[n], \quad \text{for } (P - 1 \leq n \leq L - 1),$$
$$0, \quad \text{otherwise.}$$

For the segments, the time origin $n = 0$ is defined to be at the beginning of each block as opposed to the origin of $x[n]$ or $y[n]$. Finally, for $y[n]$ we have,

$$y[n] = \sum_{r=0}^{\infty} y_r[n - r(L - P + 1) + P - 1].$$

The corresponding output modules, $y_r[n]$, for the procedure, are shown in Figure 4.4. The procedure is called the overlap-save method because the input segments overlap, so that each succeeding input section consists of $(L - P + 1)$ new points and $(P - 1)$ points saved from the previous input section [32].

4.2.2 ALGORITHM OPTIMIZATION

For greater speed and efficiency, using the block convolution approach, there are convolution algorithms that have been designed for online or real-time applications, where minimizing both input/output delay (latency) and execution time are important. Some of these algorithms [34] are simply extensions of the previously discussed overlap-save method, but provide techniques for the additional partitioning of a long impulse response $h[n]$ into J sections of equal length K (see Figure 4.5). Therefore, each impulse response partition consists of a section $h_j[n]$, of uniform length K, and a delay equal to the corresponding time-offset or $j(K - 1)$ points of that partition into the impulse response, $h[n]$. This method is called the partitioned overlap-save algorithm [34] and provides some insight into the efficiency of the algorithm driving the linear convolution module developed by Farina [16], which is the convolution module used in the present study. Furthermore, there are attempts at even greater efficiency in convolution with specific micro-algorithms that are designed to determine the optimal, nonuniform, filter partition lengths necessary for a desired output. That is, for a specified input/output delay and filter length, there are algorithms that find the nonuniform filter partition length that minimizes the computational cost of the convolution required [35].

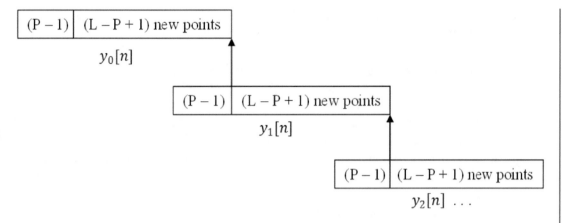

Figure 4.4: Output blocks $y_r[n]$ from the overlap-save method with each modular output producing a window of L convolved data points.

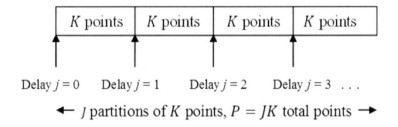

Figure 4.5: Modular blocks of the impulse response, $h[n]$.

Theoretically, it may seem that the increase in the number of arithmetic operations and memory references produced by the partitioned overlap-save algorithm would make the process less efficient than the unpartitioned overlap-save method. At the very least, all of the partitions need to be extracted, stored, recalled, filtered, and appended together, in order to produce a single, unified, modular output. Alternatively, partitioning may provide some relief on the heavy FFT calculations as one long impulse response P-point block FFT is replaced by the J shorter K-point FFTs (see Figure 4.5). However, with the reduced FFT computational load according to $N\log_2 N$, the performance increase is theoretically quite small [33].

Each partition $h_j[n]$ is treated as its own separate impulse response, and convolved by a standard overlap-save process, making use of FFT windows typically of input block length $L = 2K$, which is twice the length of the filter partition. Typically, a factor of two gives the best efficiency to the overlap-save algorithm [33]. In this fashion, the previous overlap-save condition ($P < L$) is maintained ($K < L$). Furthermore, setting the input block length, L, to twice that of the impulse response, K, is also maintained. Each K-sample section is zero-padded to the length $L = 2K$, and transformed with an FFT, so that a collection of frequency-domain filters is obtained. The results of the multiplications of these filters with the FFTs of the input blocks are appended, producing the same result as the unpartitioned convolution, by means of proper delays applied to the blocks of convolved data [34]. The primary advantage, compared to unpartitioned convolution, is that the latency of the entire filtering process is just $L = 2K$ points, which is the window of convolution for each impulse response partition and corresponding input block, as opposed to a window of $L = 2P$ points, which, in maintaining the factor of two ratio for the overlap-save method, is twice the length of the unpartitioned impulse response. In addition, an advantage of uniform partitioning is that it allows for fewer applications of the FFT, as each input block of length L requires only one FFT for all J sections, also of length $L = 2K$, of the partitioned impulse response. This is in contrast to nonuniform partitioning, where the impulse response partitions are of various lengths, and therefore, a separate FFT is performed on each input block corresponding to the window of that specific partition. That is, the convolution window of length $L = 2K$ is continually changing in accordance with the length K.

Partitioning the impulse response in order to increase the efficiency of calculating a long FFT offers some advantages. Furthermore, as the number of partitions increases, a large part of the computational load is shifted from the FFT calculations to the multiply and add steps, which are easily optimized. On modern processors, this optimization is accomplished through a few lines of hand-coded assembler instructions, such as the machine-specific, single-instruction multiple-data (SIMD) set of instructions. This instruction library allows for executing several arithmetic operations in a single assembler instruction, making the resulting code very efficient [34, 36]. Therefore, with partitioning, and producing a shorter window of data, the difficult task of FFT optimization becomes easier and more implementations become available. However, this does not mean that increasing the number of partitions to any number automatically provides a more efficient convolution algorithm. Generally, selecting a suitable number of partitions along with the type of FFT implementation is the more important task in achieving a high level of efficiency [35, 37]. The type of radix implementation provided by the FFT algorithm could also be an issue. As an example, if a finite-length N = 33,000 FFT was required, with a radix 2 implementation, a 65,536 (2^{16}) length FFT would have to be calculated as a 32,768 (2^{15}) length would not be sufficient. Consequently, the processing sequence is approximately twice the length of what is required. More recently, split-radix algorithms have been developed in an attempt to solve this problem [36].

Performance benchmarks clearly show that, if done properly, the partitioned convolution algorithm actually outperforms unpartitioned convolution in all measured cases, if a suitable number of partitions is chosen [34]. The performance comparison between the two approaches essentially becomes a comparison of the speed of the FFT implementation and the small assembler loop performing the multiply and add step. As mentioned in the previous paragraph, an additional performance consideration is the fact that a particular FFT algorithm may be optimal at certain lengths given the type of radix implementation employed. It is important to note that, for a reasonably short impulse response, of say, 33,000 points, a properly optimized ordinary overlap-save algorithm may still equal the performance of a partitioned algorithm [37]. Therefore, if sufficient, it may be more practical to simply convert the entire sequence using an appropriately efficient FFT algorithm, especially if latency is not an issue.

4.3   THE PROBLEM OF DECONVOLUTION

4.3.1   DEFINING THE PROBLEM

The equations and simplified examples of Section 4.1 illustrate that the concept of convolution is relatively straightforward. Furthermore, continuous-time or analog convolution is also a natural everyday phenomenon. However, deconvolution presents a different problem that is complex and difficult to manage. The problem of deconvolution is essentially one of separating two signals that have been convolved [38]. To help illustrate the concept, we may first apply the commutative property of the discrete-time convolution sum, that is,

$$y[n] = x[n] * h[n] = h[n] * x[n] = \sum_{k=-\infty}^{\infty} x[k]h[n-k]$$

$$= \sum_{k=-\infty}^{\infty} h[k]x[n-k]. \tag{4.6}$$

As illustrated in the examples of Section 4.1, computing $y[n]$ through the convolution sum does not present any conceptual difficulty. However, from Equation (4.6), the process of separately computing $h[n]$ or $x[n]$ from $y[n]$, that is, the process of deconvolution, does not appear to be so straightforward. The process becomes increasingly complicated when independent additive noise is present, contributing to the corruption of $y[n]$.

4.3.2   INVERSE FILTERS

The Fourier Transform Inverse Filter
One method of performing deconvolution is through the development of a linear inverse filter of one of the signals being separated. From the convolution theorem of the FT as in Equation (4.4), we have,

$$Y\left(e^{j\omega}\right) = X\left(e^{j\omega}\right) H\left(e^{j\omega}\right), \tag{4.7}$$

where X is the FT of the measurement signal, H is the transfer function or frequency response of the filter, and Y is the FT of the recorded or observed result of the LTI system. Therefore, from Equation (4.7), the deconvolution or separation of $H\left(e^{j\omega}\right)$ may be computed as

$$H\left(e^{j\omega}\right) = \frac{Y\left(e^{j\omega}\right)}{X(e^{j\omega})} = \frac{1}{X(e^{j\omega})}Y\left(e^{j\omega}\right) = P\left(e^{j\omega}\right) Y\left(e^{j\omega}\right). \tag{4.8}$$

According to Equation (4.8), in order to derive the filter response $H\left(e^{j\omega}\right)$, the filter $P\left(e^{j\omega}\right)$ is simply the inverse of the FT of the original input signal, $X(e^{j\omega})$. By designing the inverse filter, the computation is modified from that of deconvolution or division of FTs, to the increased efficiency of convolution through multiplication. Furthermore, in extending the FT approach, the problem may be shifted from one of multiplication to that of addition by taking the complex logarithms of both sides of Equation (4.7), that is,

$$\log Y(e^{j\omega}) = \log X(e^{j\omega}) + \log H(e^{j\omega}), \tag{4.9}$$

or similarly,

$$\log H(e^{j\omega}) = \log Y(e^{j\omega}) - \log X(e^{j\omega}).$$

There are different methods, other than the FT approach, such as matrix formulation, that can be used for designing an inverse filter. The method presently discussed uses the FT and its associated properties. However, there are limitations to this approach.

Initially, the FT approach requires that a signal has a respective FT. Primarily, absolute summability is a sufficient condition for the existence of an FT representation, and it also guarantees uniform convergence [10]. As an example, let an input exponential sequence be defined as $x[n] = a^n u[n]$. The FT of this sequence is [32]

$$X(e^{j\omega}) = \sum_{n=0}^{\infty} a^n e^{-j\omega n} = \sum_{n=0}^{\infty} (ae^{-j\omega})^n$$

$$= \frac{1}{1 - ae^{-j\omega}} \quad \text{if} \quad |ae^{-j\omega}| < 1 \quad \text{or} \quad |a| < 1. \tag{4.10}$$

Therefore, the condition $|a| < 1$ is the condition of absolute summability for $x[n]$. The importance of this result becomes clear if we substitute Equation (4.10) into Equation (4.8). In order for the inverse of $X(e^{j\omega})$ to exist, it must not have any zeros (roots of the numerator) or any poles (roots of the denominator) within the region or window of convergence.

Another problem with the FT inverse filter approach is the existence of additive random noise, $N(e^{j\omega})$, in the original convolution result. A more realistic representation would be an extension of Equation (4.7) as

$$Y\left(e^{j\omega}\right) = X\left(e^{j\omega}\right) H\left(e^{j\omega}\right) + N\left(e^{j\omega}\right). \tag{4.11}$$

Therefore, Equation (4.8) becomes,

$$H\left(e^{j\omega}\right) = \frac{Y\left(e^{j\omega}\right)}{X(e^{j\omega})} - \frac{N(e^{j\omega})}{X(e^{j\omega})}. \tag{4.12}$$

The nature of the signal being inverted then becomes important. As an example, in order to restore the original signal $X\left(e^{j\omega}\right)$, such as with biomedical image processing, it is sometimes necessary to derive the inverse filter of the "smearing" function $H\left(e^{j\omega}\right)$. That is,

$$X\left(e^{j\omega}\right) = \frac{Y\left(e^{j\omega}\right)}{H(e^{j\omega})} - \frac{N(e^{j\omega})}{H(e^{j\omega})}. \tag{4.13}$$

Problems arise because $H\left(e^{j\omega}\right)$ is usually a lowpass function [38], due to the frequency response of the imaging instrumentation, whereas $N\left(e^{j\omega}\right)$ is uniformly distributed over the entire spectrum. Therefore, the amplified noise at higher frequencies (the second component in Equation (4.13)) overshadows the restored image. Obviously, the same issue holds true if $X\left(e^{j\omega}\right)$ is being inverted, as in Equation (4.12), and contains poor or suppressed high-frequency content. Furthermore, where the frequencies are strongly suppressed, the SNR is poor, and the inverse function will amplify mainly the noise [38].

Consequently, in order to avoid excessive noise and a corrupted, inaccurate result, the sequence to be inverted is usually band-limited ($\omega < \omega_L$). An attempt to recover frequencies outside this band or constraint would only serve to amplify the noise unduly without retrieving informative signals. However, the frequency response may then be compromised, which affects the range and accuracy of the deconvolution result. This effect is formally known as ill-conditioning [39].

Limitations and Optimization

In order to perform deconvolution using the inverse filter process, one initial concern is how much information is available about the two convolved signals $x(t)$ or $h(t)$ in their original states. There are essentially two forms of the problem. The first and simpler problem is if one of the signals is completely known beforehand. Its inverse filter may then be carefully determined and applied to Equation (4.8) through convolution. However, in areas such as telecommunications and biomedical image processing, where deconvolution is used in an attempt to remove system response distortions in order to restore the original input, it is common to know nothing about either of the two original convolved signals. This produces the more difficult second problem of deconvolving two signals when both are unknown: this is called blind deconvolution [38].

Blind deconvolution is essentially reduced to either estimating or eliminating one of the two unknown signals. The specific algorithm or recovery approach developed largely depends on the existing information available. One approach is to determine a distortion-free model of the original input or object function and subtract it from the recorded or distorted output, leaving behind an estimate of the distorting system impulse response [38]. In considering the effects of additive noise and ill-conditioning, constraints are imposed and care is taken to confine the frequency response of the compensating filter or model to the frequency band in which the distorted output has appreciable components. In some cases, this may be achieved simply by truncating the frequency response of the resulting compensating filter [38].

4.3.3   THE WIENER FILTER

Even though its application is limited to linear, stationary, and stochastic (random) processes, the Wiener filter provides a method for optimal filtering by taking into account the statistical characteristics of signal processes. The Wiener filter, in this case, represented as $W(f)$, is designed to minimize the mean-squared error (MSE) between the output of the estimated input, $\hat{X}(f)$, and the unaffected, unknown, original input, $X(f)$. In biomedical image recovery, the Wiener filter can be used to produce an estimate of the original, unaffected image. It is also an optimal filter in the sense that no better linear filter can be found for noise reduction alone, provided that we are restricted to the knowledge that the noise is additive and Gaussian distributed [40].

A Wiener deconvolution filter attempts to minimize the impact of deconvolved noise at frequencies that have poor SNR. The method behind the Wiener filter is to reduce the amount of additive noise in a signal by comparison with an estimate of the desired noiseless signal. Therefore, the goal is to design an LTI filter, based on the comparative minimum mean-squared error (MMSE) criterion, that produces an output that is as close to the original signal as possible. The comparative MSE may be expressed as

$$\epsilon(f) = E|X(f) - \hat{X}(f)|^2, \qquad f = \frac{\omega}{2\pi}, \tag{4.14}$$

where

- E is the expectation value of a statistical random variable,

- $X(f)$ is the FT of the unknown input, and

- $\hat{X}(f)$ is the FT of the estimated input.

To illustrate how the Wiener filter may be used in deconvolution, suppose we have an LTI system that is defined by the frequency-domain relationship in Equation (4.11), but stated in the time domain as

$$y(t) = x(t) * h(t) + n(t), \tag{4.15}$$

where

- $x(t)$ is the unknown input at time, t,

- $h(t)$ is the known system impulse response,

- $n(t)$ is the known or modeled (Gaussian) additive noise, independent of $x(t)$, and

- $y(t)$ is the known system output.

The purpose of Wiener deconvolution is to determine the convolution filter, $w(t)$, in order to find an estimate, defined as $\hat{x}(t)$, of the original input $x(t)$, that minimizes the MSE defined in Equation (4.14); that is,

$$\hat{x}(t) = w(t) * y(t). \tag{4.16}$$

Applying the convolution theorem, the FT of the estimate is then

$$\hat{X}(f) = W(f)Y(f). \tag{4.17}$$

The Wiener filter itself is given by the expression [39, 40]

$$W(f) = \left[\frac{H^*(f)}{|H(f)|^2 + \frac{S_n(f)}{S_x(f)}}\right], \tag{4.18}$$

where

- $H(f)$ is the FT of the system response,

- $S_n(f)$ is the power spectral density (PSD) of the noise, $n(t)$, and

- $S_x(f)$ is the PSD of the input, $x(t)$.

We can also write

$$W(f) = \frac{1}{H(f)}\left[\frac{|H(f)|^2}{|H(f)|^2 + \frac{S_n(f)}{S_x(f)}}\right] \tag{4.19}$$

$$= \frac{1}{H(f)}\left[\frac{|H(f)|^2}{|H(f)|^2 + \frac{1}{SNR(f)}}\right], \tag{4.20}$$

where

- $SNR(f) = S_x(f)/S_n(f)$ is the SNR as a function of the PSD.

We can now make some observations regarding the Wiener deconvolution filter:

- The Wiener filter transfer function as in Equation (4.18) depends upon the PSD of the original signal and the PSD of the additive noise. As mentioned previously, the original input is not known, and therefore, if the noise is also unknown, each respective PSD must be estimated from previously determined information or models.

- When there is no additive noise, the PSD of the noise is $S_n(f) = 0$, and the Wiener filter becomes the inverse filter, $\frac{1}{H(f)}$.

- The gain of the Wiener filter is essentially modulated by the noise-to-signal PSD ratio in the denominator, as shown in Equation (4.19). If the SNR is high, the filter is close to the inverse filter, which is readily seen from Equation (4.20).

- If the noise-to-signal PSD ratio is not available as a function of frequency f, it may be set equal to a constant $K = \frac{1}{SNR}$. The filter is then known as the parametric Wiener filter [39].

- The Wiener filter is not often singular or ill-conditioned; wherever $H(f) = 0$, the estimated output $\hat{X}(f) = 0$ [38].

- As the noise at certain frequencies increases, the SNR decreases, and the term in the square brackets of Equation (4.20) also decreases. Therefore, the Wiener filter attenuates frequencies depending on their SNR.

4.3.4 DIFFICULTIES WITH DECONVOLUTION

Many applications of deconvolution involve the recovery of the original input (or object) from a composite filtered result. The opportunity for success in solving for the original input $x[n]$ exists only if the solution of $x[n]$ actually exists. Furthermore, adding noise to the original signal produces an environment of corruption that may result in a problem that has no solution, is ill-posed, or is at best, ill-conditioned [31].

A problem is ill-posed when at least one of three conditions exist:

- there is no solution,

- a solution exists but is not unique, or

- its solution does not depend continuously on the data [31].

Even if the problem passes all three tests for well-posedness, a proper solution may depend sensitively on small fluctuations in the data. The difficulty with identifying these fluctuations increases significantly if noise is present. Furthermore, because all data have uncertainty, such sensitivity limits the reliability and value of a solution. Problems exhibiting this behavior are said to be ill-conditioned [31]. Therefore, a series of constraints or processing limitations need to be applied to the problem in order to provide a reasonable or useful result.

4.4 REMARKS

Constraints are important in performing efficient deconvolution. Therefore, it is important to determine carefully the parameters of a constraint, and evaluate how these parameters will contribute to the accuracy of the result. Limiting the bandwidth of a deconvolution procedure (or filter) is an example of a constraint. A bandwidth constraint that is too narrow may cause a "choking" limitation on the frequency response of the result. However, a bandwidth that is too wide may result in unwanted signal distortion due to the presence of amplified additive noise.

The next chapter examines specific filtering techniques that have been used in the present work in order to increase the efficiency and accuracy of acoustic measurement using a source loudspeaker. One of these techniques uses deconvolution to remove and filter out the effects of the transducers on the measured AIRF. That is, in order to remove the effects of the reference loudspeaker and recording microphones from the measurement process, their collective impulse response is first measured separately in an anechoic environment. A proper inverse filter of this response, which includes a constrained bandwidth in order to achieve a suitable deconvolution, is then determined and subsequently applied to the measured AIRF. This method is expected to provide an improved representative AIRF measurement as the effects of the measurement hardware are removed from the process.

CHAPTER 5

Experimental Method for the Derivation of an AIRF of a Music Performance Hall

5.1 GENERAL TECHNICAL SPECIFICATIONS

ESS measurement signal generation, recording, and post-processing were performed digitally with a 2.0 GHz MacBook laptop computer, using the Adobe Audition 2.0 recording software package, operating at 96 kHz and 32 bits/sample. Additional signal processing, including the derivation of the ESS measurement signal, its inverse filter, and all convolutions were performed using the AURORA suite of signal processing modules developed by Farina [16]. The self-powered Meyer MTS-4 HF horn loudspeaker system [23] was used as the reference source for all measurement signals. Recordings were captured with the Studio Projects C3 condenser microphone [41]. Signal conversions between analog and digital were performed with the MOTU 896HD FireWire computer interface operating at 96 kHz and 24 bits/sample. The signal path for generating and recording the ESS is presented in Figure 5.1.

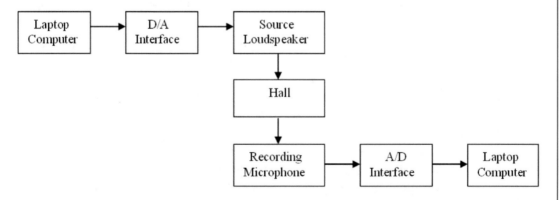

Figure 5.1: Block diagram for generating and recording the ESS.

5.2   MEASUREMENT RECORDING CONFIGURATIONS

There are many types of microphones that may be considered for recording music or measuring the acoustical characteristics of a hall. The differences between microphones are dependent primarily on two criteria: first, the microphone classification, that is, if it is a dynamic, ribbon, or condenser type; and second, the polar response or pickup pattern of the microphone. For classical music recording, where a low level of system noise is important, the preferred microphones are condenser or ribbon types with a wide, flat frequency response, and very low self-noise. In these applications, a microphone A-weighted self-noise specification of less than 21 dB equivalent SPL is recommended [42]. Microphones with directional polar responses, such as cardioids (heart-shaped), are designed to capture the maximum amount of acoustic energy provided in a particular direction of arrival. Microphones with cardioid patterns have maximum acceptance at 0 degrees (on axis) and maximum rejection at 180 degrees (off axis). One advantage with this type of microphone (pattern) is that it gives a better sense of directivity and has the ability to reject unwanted sounds arriving off axis. This feature is very important in live sound reinforcement where the rejection capability of a microphone often determines clarity of sound and elimination of feedback. Omnidirectional microphones capture a maximum amount of acoustic energy equally over a polar response of 360 degrees. Because these omnidirectional patterns do not have a point of maximum rejection, they consequently provide little sense of direction. Due to this feature, they are rarely used in live sound reinforcement. However, when used in recording or acoustic measurement, they may provide a better sonic representation of the venue or hall, as they are designed to capture sound equally from the sides and rear, as well as the front.

In the present work, mono and stereo recordings were performed using the large-diaphragm, multi-pattern, Studio Projects C3 condenser microphone. The multi-pattern feature represents switchable polar response control between cardioid, omnidirectional, and figure-of-eight. In addition, the C3 provides an acceptable A-weighted self-noise specification of 18 dB equivalent SPL [41]. Each C3 microphone is accompanied by its own VTB1 analog preamplifier, and for digital recording purposes, was linked to a MOTU 896HD 24-bit IEEE 1394 FireWire interface. Each microphone output was monitored using a combination of VTB1 preamplifier volume-unit (VU) metering and the front-panel 10-segment peak-reading light-emitting-diode (LED) metering on the 896HD.

For the purposes of acoustical measurement, most temporal parameters are computed from a monaural impulse response captured with an omnidirectional microphone [17, 18]. This is because, with a single microphone, the omnidirectional polar response provides a better overall sonic representation of the room, as it accepts reflections from all directions. Figure 5.2 represents a standard mono-to-mono configuration, where the microphone (receiver) is placed directly on-axis (0 degree) in front of the measurement loudspeaker (source). In order to maintain flexibility of analysis, the present study performed mono ESS recordings using both omnidirectional and cardioid polar responses.

Recording the temporal and spatial stereophonic propagation of a mono source is meant to simulate the way a human being perceives sound. Figure 5.3 represents a standard mono-to-

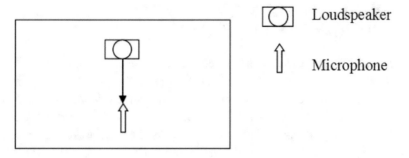

Figure 5.2: Mono recording.

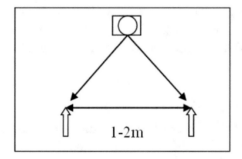

Figure 5.3: Stereo spaced-pair or AB recording.

stereo spaced-pair recording configuration. With the spaced-pair, or AB technique, two identical microphones are placed in parallel, aiming straight ahead toward the musical ensemble. The AB technique essentially relies on time-of-arrival differences between microphones for spatial image placement in the stereo field. An increase in spacing produces a wider stereo spread. However, if the distance between the microphones is too great, the resulting stereo image will be exaggerated. Consequently, the spacing of the microphones is variable and partly determined by the size of the sound source; that is, a symphony orchestra will require a different spacing than a single performer. For music recording applications, the proper distance is often simply determined through listening on headphones, in order to ensure that the stereo imaging is acceptable. For ESS recording applications, the two microphones should be spaced approximately 1.5 meters from each other [43]. Indeed, with this recording configuration, the omnidirectional pattern is the most popular. However, once again, for purposes of flexibility, the present study performed recordings with both cardioid and omnidirectional patterns.

5.3   FILTERING AND REMOVAL OF THE EFFECTS OF THE TRANSDUCERS ON THE AIRF

The derivation of a faithful and accurate AIRF requires that the effects of the transducers used in the measurement process are properly managed and filtered. In the present work, it is shown that, first, by performing a pink noise spectral analysis of the measurement loudspeaker beforehand, in a separate session and a suitable "reflection-free" environment, it is possible to develop a loudspeaker equalization prefilter, which will ensure that any output from the loudspeaker possesses a spectrum similar to that of the input. In this manner, transparency is maintained as the effect of the loudspeaker is virtually removed from the process. This is important toward achieving complete isolation of the AIRF.

In addition, during the same session as above, a separate impulse response analysis was performed for the measurement transducer system. From this impulse response, a transducer inverse filter was then developed and applied to the final process. With this technique, the derived AIRF of the hall is further isolated from the effects of the electromechanical transducers or measurement system for accurate documentation.

5.3.1   LOUDSPEAKER EQUALIZATION PREFILTER

Pink noise spectral analysis of the Meyer MTS-4 measurement loudspeaker was performed in a separate session and in an acceptable "reflection-free" outdoor location, with the resulting digital equalization curve (Adobe Audition 2.0 graphic equalizer) being stored as a pre-insert for the measurement loudspeaker. These types of real-time-analysis (RTA) curves are performed and designed to ensure that the frequency response of the loudspeaker is uniform or "flat." This process helps to maintain transparency of the system as the frequency response of the loudspeaker output is similar to that of the input when ESS measurements are performed (see Figure 5.4).

An appropriate anechoic chamber, with a frequency response of approximately 20 Hz to 20 kHz, was not available in Calgary. Therefore, an acceptable outdoor location for the anechoic environment was determined to be Lot 3A behind the Eckhardt-Gramatté Hall. In order to further improve measurement conditions, the measurement loudspeaker was supported on its back, 30 cm off the ground, pointing upward. The Earthworks M23 measurement microphone was placed 1.7 m over the loudspeaker. In addition, the measurements were performed late at night to ensure that any environmental disturbances were kept to an absolute minimum level.

5.3.2   TRANSDUCER IMPULSE RESPONSE AND INVERSE FILTER

In the present study, the goal of deriving the AIRF is to isolate and analyze the acoustical transfer function of the hall without the effects produced by the necessary transducers such as the microphones, amplifiers, and loudspeakers. A primary advantage of the ESS method is that it allows the loudspeaker harmonic distortion products to be identified and edited after convolution has been performed (as shown in Figures 3.6 and 3.7). However, there are other electro-acoustical effects

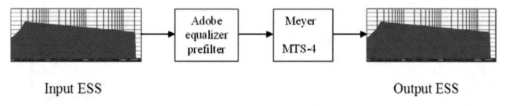

Input ESS   Output ESS

Figure 5.4: Meyer MTS-4 loudspeaker equalization prefilter, designed to produce a loudspeaker output spectrum that is correspondingly similar to the respective input spectrum.

(residual coloring) produced by the amplifier, loudspeaker, and microphone combination that affect the linear impulse response and are more difficult to edit or remove.

ESS derivation of the impulse response of the transducer system was performed separately during the same session and in the same environment that produced the loudspeaker prefilter. The time-domain representation of the transducer system impulse response and the corresponding transducer system inverse filter (TSIF) are shown in Figures 5.5 and 5.6. The derivation of the inverse filter was performed using the Kirkeby inverse filter module in AURORA [16].

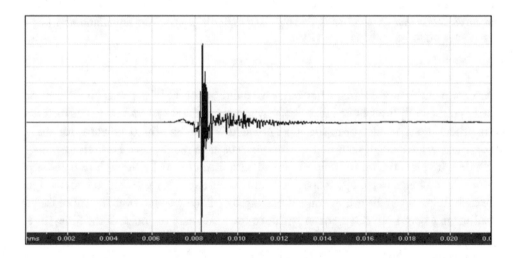

Figure 5.5: Measured impulse response of the ESS measurement system. Time-scale: 2 ms per division; amplitude-scale: 3 dB per division. Reproduced, with kind permission from IEEE, from D. Frey, V. Coelho, and R.M. Rangayyan, "Filtering and removal of the effects of the transducers on the acoustical impulse response of concert halls," *IEEE Canadian Conference on Electrical and Computer Engineering (CCECE)*, St. John's, Newfoundland and Labrador, Canada, 3–6 May, 2009, pp. 368–371. © IEEE.

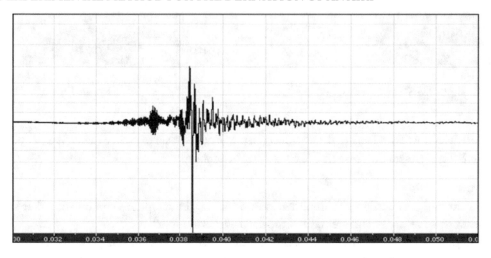

Figure 5.6: Result of the ESS measurement system inverse filter produced by the Kirkeby inverse filter module in AURORA [16]. Time-scale: 2 ms per division; amplitude-scale: 3 dB per division. Reproduced, with kind permission from IEEE, from D. Frey, V. Coelho, and R.M. Rangayyan, "Filtering and removal of the effects of the transducers on the acoustical impulse response of concert halls," *IEEE Canadian Conference on Electrical and Computer Engineering (CCECE),* St. John's, Newfoundland and Labrador, Canada, 3–6 May, 2009, pp. 368–371. © IEEE.

The choice of the "usable" frequency range for building the TSIF depends on the transducers involved. It is difficult to filter or equalize properly a measurement system outside of the effective bandwidth where the transducers are the most efficient. Therefore, the bandwidth for the TSIF depends on the specific loudspeakers and microphones used in the process. Furthermore, the filter bandwidth that produces the most useful result is determined essentially through trial and error. If an excessively broad frequency range is specified, and because it is not possible to boost a signal at extremely high frequencies, the average signal magnitude will be lowered over the entire frequency range, resulting in a poor SNR. Consequently, due to the increase in the noise of the filter, the result of the subsequent deconvolution may become increasingly corrupted. In the present study, through much experimentation, the most efficient bandwidth of the Kirkeby inverse filter was determined to be from 100 Hz to 11 kHz.

When the measured transducer or "anechoic" impulse response is convolved with the corresponding TSIF, the result is an approximation of the Dirac delta function (see Figure 5.7); this illustrates that the input signals are inverses of each other. Although the TSIF does not provide an ideal Dirac impulse function, it does provide a useful result. This is illustrated by the frequency-domain or spectral representations of the original impulse response of the transducers by itself followed by the same impulse response after convolution with the derived TSIF (see Figures 5.8 and

5.9). The TSIF is effective in providing a more uniform transfer function. In this manner, because the effects of the transducers can now be removed from the measured AIRF of a music performance hall, they may be employed in the ESS process without any significant biasing effect.

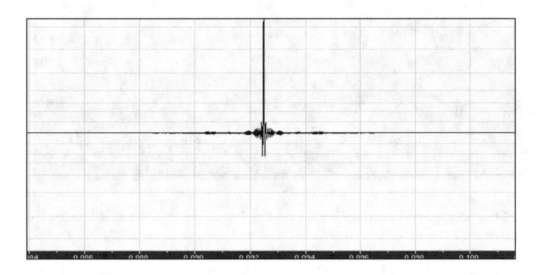

Figure 5.7: Result of convolution of the reference transducer impulse response (Figure 5.5) and the corresponding inverse filter (Figure 5.6). Time-scale: 2 ms per division; amplitude-scale: 3 dB per division. Reproduced, with kind permission from IEEE, from D. Frey, V. Coelho, and R.M. Rangayyan, "Filtering and removal of the effects of the transducers on the acoustical impulse response of concert halls," *IEEE Canadian Conference on Electrical and Computer Engineering (CCECE)*, St.John's, Newfoundland and Labrador, Canada, 3–6 May, 2009, pp. 368–371. © IEEE.

However, the following question arises: Where in the derived AIRF process is it best to apply the transducer inverse filter? Essentially, it may be applied to the ESS signal prior to the measurement loudspeaker when recording in the hall, or later in post-processing, either to the recorded sweep before deconvolution, or after, to the derived AIRF. Applying the filter to the test signal prior to the loudspeaker will likely produce a weaker ESS signal for recording. Furthermore, it may also produce signal distortion in the playback step as the loudspeaker prefilter (see Figure 5.4) has also been inserted prior to the loudspeaker. Therefore, it is advisable to apply the transducer inverse filter in the post-processing stage. It makes the most sense to apply the transducer inverse filter to the final edited AIRF (see Figure 5.10); this is because the derived AIRF is much shorter in duration as compared to the recorded 15-second ESS. Consequently, the computation required to perform the convolution with the transducer inverse filter is reduced.

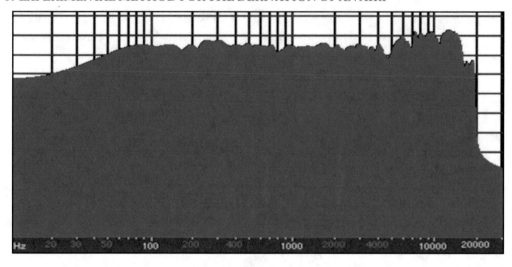

Figure 5.8: Spectrum of the transducer impulse response. Frequency range: 0 Hz to 30 kHz (log scale); amplitude-scale: 12 dB per division. Reproduced, with kind permission from IEEE, from D. Frey, V. Coelho, and R.M. Rangayyan, "Filtering and removal of the effects of the transducers on the acoustical impulse response of concert halls," *IEEE Canadian Conference on Electrical and Computer Engineering (CCECE)*, St. John's, Newfoundland and Labrador, Canada, 3–6 May, 2009, pp. 368–371. © IEEE.

5.4   VERIFICATION OF THE AIRF REPRESENTATION

The development of the transducer inverse filter (Section 5.3.2) and the evaluation of software algorithm performance using pink noise (Section 2.2) have shown that frequency response analysis may be used to monitor and maintain processing accuracy at various stages of the acoustic measurement process. In addition, frequency response analysis may be used to provide verification of the experimentally derived AIRF of a music performance hall; that is, through the use of a common test signal, spectral analysis of the AIRF of the actual hall may be performed and compared with that of the experimentally derived AIRF.

Verification of representation is a difficult problem and a primary concern in the derivation of an AIRF. The sound quality of the AIRF is important only in that it properly represents the hall being characterized. In other words, the fact that an AIRF sounds good does not necessarily mean that it accurately represents the hall from which it was derived. In practice, with the different acoustical impulse responses or "convolution reverbs" available on the market, it is difficult to assess the measurement accuracy of the data provided and to determine whether or not they provide faithful representations of the acoustic spaces in which they were measured.

With the ESS technique of measuring an AIRF, one existing method of measurement verification is to use a common, suitable test signal and conduct formal listening tests of the derived

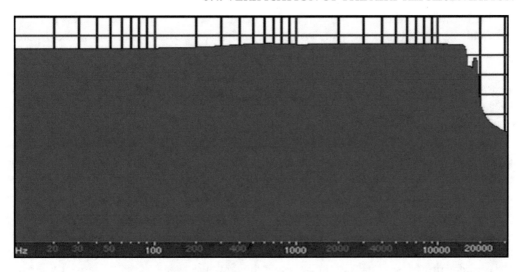

Figure 5.9: Spectrum of the result of convolution shown in Figure 5.7. Frequency range: 0 Hz to 30 kHz (log scale); amplitude-scale: 12 dB per division. Reproduced, with kind permission from IEEE, from D. Frey, V. Coelho, and R.M. Rangayyan, "Filtering and removal of the effects of the transducers on the acoustical impulse response of concert halls," *IEEE Canadian Conference on Electrical and Computer Engineering (CCECE)*, St.John's, Newfoundland and Labrador, Canada, 3–6 May, 2009, pp. 368–371. © IEEE.

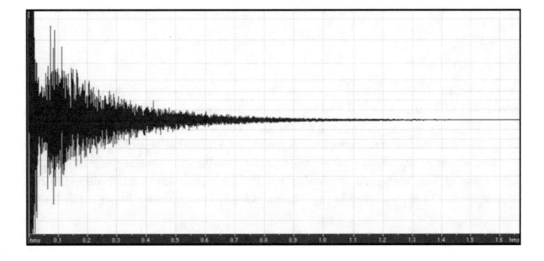

Figure 5.10: Final edited AIRF after convolution with the transducer inverse filter. Time-scale: 0.1 s per division; amplitude-scale: 3 dB per division.

AIRF with that of a recording performed in the hall itself. A more accurate and less subjective method would be to perform a comparative spectral analysis of the two impulse responses in order to determine their similarity. However, this is not easily accomplished since the effects of the transducers need to be separately measured and then removed from each of the two processes so that the resulting impulse responses, actual and derived, can be isolated for analysis. Furthermore, there are the problems associated with deconvolution and the process of removing the common test signal from the recording in order to isolate the AIRF of the real hall.

Spectral comparisons of our experimentally derived AIRF with the recorded AIRF of the actual hall were performed using a software-generated, dry, ceramic drum loop as the common test signal (the term "dry" indicates a synthesized signal without the effects of any transducers or hall). The ceramic drum test signal was generated for recording through the same reference Meyer MTS-4 loudspeaker system that was used for the AIRF derivation process. Because the measurement transducer system is identical for each drum signal hall recording and the respective AIRF derivation, the transducer inverse filter obtained as previously discussed can be applied to both processes, leaving only the common drum test signal and the corresponding impulse response, real or derived; therefore, spectral comparisons can be performed. Apple Logic 8 Space Designer convolution software was used for real-time playback convolution of the derived AIRF with the drum signal. Figure 5.11 illustrates the spectrum of the dry, ceramic drum test signal.

Figure 5.12 provides a spectral comparison of the loudspeaker-generated drum test signal recording performed in the Eckhardt-Gramatté Hall (red line) and the drum test signal convolved with the experimentally derived AIRF of the same hall (solid blue region). Within the bandwidth of 100 Hz to 11 kHz (the specified range of the transducer inverse filter), the result of convolution with the derived AIRF provides a reasonably faithful representation of the drum recording performed in the actual hall.

5.5   REMARKS

Measurement verification is important in the assessment of the derivation of an AIRF of a music performance hall. Objective numerical verification is the ultimate goal as listening tests by themselves provide only subjective analysis and qualitative results. Spectral analysis as illustrated in Section 5.4 provides a basis for numerical verification and comparison of a direct hall recording and convolution of the corresponding derived AIRF. The bandwidth of the transducer inverse filter plays an important role in the range of accuracy within the final AIRF. For the present study, AIRF derivations and spectral comparisons were performed in three different halls. The results are presented in the next chapter.

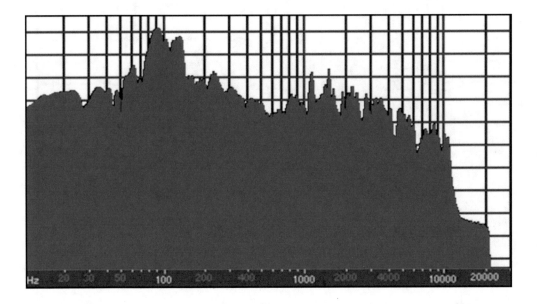

Figure 5.11: Spectrum of software-generated, dry, ceramic drum test signal. Frequency-scale: 0 Hz to 30 kHz (log scale); amplitude-scale: 12 dB per division. Reproduced, with kind permission from IEEE, from D. Frey, V. Coelho, and R.M. Rangayyan, "Spectral verification of an experimentally derived acoustical impulse response function of a music performance hall," *IEEE Canadian Conference on Electrical and Computer Engineering (CCECE)*, Calgary, Alberta, Canada, 2–5 May, 2010, pp. 1–4. © IEEE.

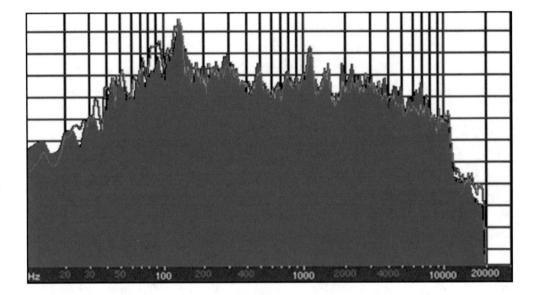

Figure 5.12: Spectrum of the result of convolution of the dry, ceramic drum test signal with the derived AIRF (solid blue region) of the Eckhardt-Gramatté Hall. The spectrum of the actual hall drum signal recording is shown by the red line. Frequency-scale: 0 Hz to 30 kHz (log scale); amplitude-scale: 12 dB per division. Reproduced, with kind permission from IEEE, from D. Frey, V. Coelho, and R.M. Rangayyan, "Spectral verification of an experimentally derived acoustical impulse response function of a music performance hall," *IEEE Canadian Conference on Electrical and Computer Engineering (CCECE)*, Calgary, Alberta, Canada, 2–5 May, 2010, pp. 1–4. © IEEE.

CHAPTER 6

Evaluation of Results

Mono and stereo ESS recordings were performed in three music performance halls: the Eckhardt-Gramatté Chamber Music Hall, the Husky Oil Great Hall, and the Rehearsal Hall, all located in the Rozsa Centre at the University of Calgary. The unedited mono-recorded sweeps are illustrated in Figure 6.1. Readily apparent are the differences in the low-frequency responses (below 100 Hz) of the 15-second, 22 Hz–22 kHz ESS. In particular, the response of the Great Hall in Figure 6.1(b) has a gradually rising, more controlled low-frequency response, free of unwanted resonant peaks. In comparison, the low-frequency response of the Eckhardt-Gramatté Hall in Figure 6.1(a) has a 51-Hz resonant peak appearing just after the 2-s mark. This low-frequency resonance is not unexpected as the audience seating area in the Eckhardt-Gramatté Hall is built on a suspended floor in order to accommodate the air conditioning system under it. Consequently, it was observed that this cavity-like structure acts as a low-frequency resonant chamber. Furthermore, through spectral analysis of the hall, the low-frequency band of 40–60 Hz was measured to be naturally excessive. It is interesting to note that these features in the low-frequency responses of the different halls are confirmed through their final derived AIRF aural representations. That is, upon post-production listening to the convolution reverb of each hall, it was the Great Hall (illustrated in Figure 6.1(b)) that possessed the most pleasing and uniform low-frequency response.

Using the "convolve with clipboard" convolution module in AURORA [16], each mono ESS recording shown in Figure 6.1 was convolved with the inverse filter of the original ESS measurement signal. The unedited results of the deconvolution, showing both linear and nonlinear responses, are illustrated in Figure 6.2. In each deconvolution result, the primary response is the desired linear impulse response of the hall. Preceding the primary response are the increasing orders—from right to left—of harmonic distortion, each represented by its own impulse response [8, 15]. Representation in this manner demonstrates interesting nonlinear behavior of the reference loudspeaker system, while allowing for easy separation, editing, and documentation of the desired linear response.

A primary difference between the results of the deconvolution in Figure 6.2 is the level of the second harmonic distortion component with the larger Eckhardt-Gramatté Hall (384 seats). This is probably due to the higher SPL (112 dB at 1.7 m) through the compression driver and horn unit of the Meyer MTS-4 reference loudspeaker. High pressures, if propagated through a horn with little reduction in level, will generate high amounts of second harmonic distortion [28]. Comparatively, the two smaller halls each measured a lower SPL of 99 dB (at 1.7 m) through the same reference loudspeaker. Consequently, their levels of second harmonic distortion are correspondingly lower in magnitude. Ultimately, however, the nonlinear distortion components of the reference loudspeaker are edited and removed from the final AIRF. This is one advantage of using the ESS method

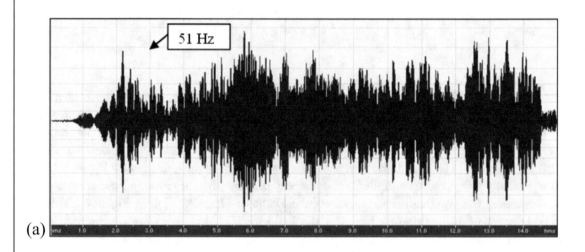

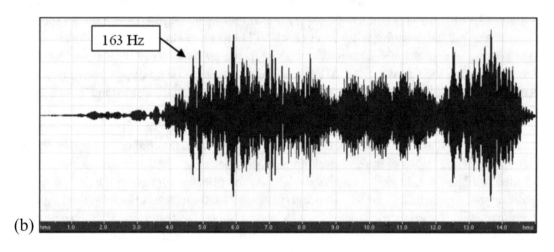

Figure 6.1: Unedited 15-second mono ESS recordings from: (a) the Eckhardt-Gramatté Hall; (b) the Husky Oil Great Hall; and (c) the Rehearsal Hall. Time-scale: 1 s per division; amplitude-scale: 3 dB per division. Reproduced, with kind permission from IEEE, from D. Frey, V. Coelho, and R.M. Rangayyan, "Filtering and removal of the effects of the transducers on the acoustical impulse response of concert halls," *IEEE Canadian Conference on Electrical and Computer Engineering (CCECE)*, St. John's, Newfoundland and Labrador, Canada, 3–6 May, 2009, pp. 368–371. © IEEE. *(Continues.)*

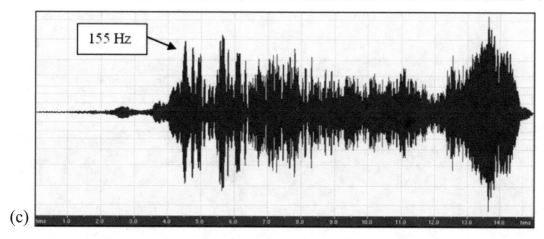

Figure 6.1: *(Continued.)* Unedited 15-second mono ESS recordings from: (a) the Eckhardt-Gramatté Hall; (b) the Husky Oil Great Hall; and (c) the Rehearsal Hall. Time-scale: 1 s per division; amplitude-scale: 3 dB per division. Reproduced, with kind permission from IEEE, from D. Frey, V. Coelho, and R.M. Rangayyan, "Filtering and removal of the effects of the transducers on the acoustical impulse response of concert halls," *IEEE Canadian Conference on Electrical and Computer Engineering (CCECE)*, St. John's, Newfoundland and Labrador, Canada, 3–6 May, 2009, pp. 368–371. © IEEE.

of acoustic measurement as it enables high levels of loudspeaker SPL in order to achieve proper stimulation of the hall for recording, post-processing, and documentation.

6.1   THE EDITED AIRF DERIVATIONS OF THE THREE MUSIC PERFORMANCE HALLS

The length of time it takes for the level of a sound to decay by 60 dB is defined as the reverberation time (RT60) of a room [44]. In order to edit a derived AIRF to its proper length, it is necessary to measure separately the RT60 of the hall beforehand. The calculation of an RT60 has its own set of standard measurement parameters and procedures [18], including multiple-octave-band calculations used for final averaging. Multi-band measurements are important due to the fact that, in any room, different frequencies will possess different decay times depending on various acoustic parameters, such as absorption coefficients for certain materials at specific frequencies. The six measurement octaves are typically 125, 250, 500, 1,000, 2,000, and 4,000 Hz [18]. For the present study, the acoustician of the Eckhardt-Gramatté Hall, N. Jordan (of Jordan Akustik), was consulted on his particular method for measuring the RT60. Jordan Akustik [45] uses its own highly developed Odeon room acoustic modeling software for calculating reverberation time. The method essentially consists of two speaker positions on stage, one forward and one back to the side, along with eight recording

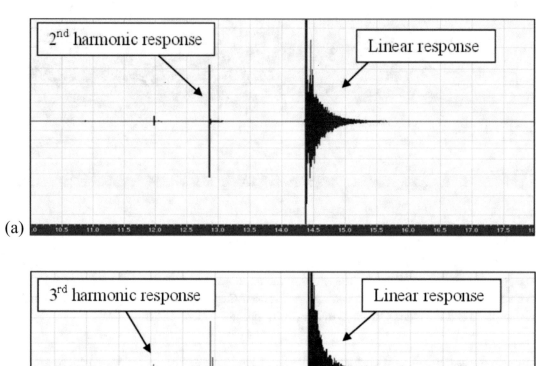

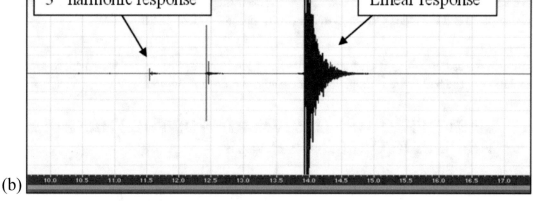

Figure 6.2: Unedited results of deconvolution for: (a) the Eckhardt-Gramatté Hall; (b) the Husky Oil Great Hall; and (c) the Rehearsal Hall. Time-scale: 0.5 s per division; amplitude-scale: 3 dB per division. Reproduced, with kind permission from IEEE, from D. Frey, V. Coelho, and R.M. Rangayyan, "Filtering and removal of the effects of the transducers on the acoustical impulse response of concert halls," *IEEE Canadian Conference on Electrical and Computer Engineering (CCECE)*, St.John's, Newfoundland and Labrador, Canada, 3–6 May, 2009, pp. 368–371. © IEEE. *(Continues.)*

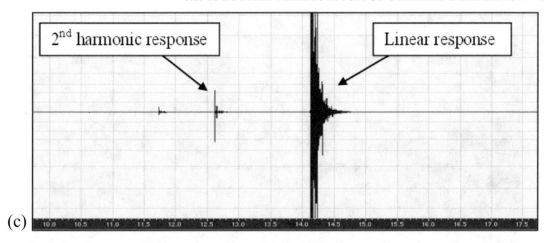

(c)

Figure 6.2: *(Continued.)* Unedited results of deconvolution for: (a) the Eckhardt-Gramatté Hall; (b) the Husky Oil Great Hall; and (c) the Rehearsal Hall. Time-scale: 0.5 s per division; amplitude-scale: 3 dB per division. Reproduced, with kind permission from IEEE, from D. Frey, V. Coelho, and R.M. Rangayyan, "Filtering and removal of the effects of the transducers on the acoustical impulse response of concert halls," *IEEE Canadian Conference on Electrical and Computer Engineering (CCECE)*, St.John's, Newfoundland and Labrador, Canada, 3–6 May, 2009, pp. 368–371. © IEEE. *(Continues.)*

positions used over the audience area. Using pink noise as an impulse measurement source, the multi-band decay function (in Odeon) computes the RT60 for the six octave bands simultaneously. The final RT60 value is then the average of the six main octave bands recorded at each position from each loudspeaker. Pure tones are not acceptable for measurement signals as they are too specific with a limited bandwidth. With approval from Mr. Jordan, his measurement results of the RT60 from the Eckhardt-Gramatté Hall (1.66 seconds) were used for this study. Furthermore, these same results were used to approximate the RT60s of the smaller Husky Oil Great Hall and Rehearsal Hall. The final edited AIRF derivations for the three music performance halls are illustrated in Figure 6.3. It is interesting to note the response of the Rehearsal Hall in that it contains more high-level, delayed reflections producing an increasingly distorted response compared to the other two halls. The Rehearsal Hall is a small, rectangular room with hard surfaces, which causes the reflections noted above.

6.2   SPECTRAL VERIFICATION OF THE AIRF DERIVATION

Figure 6.4 provides spectral comparisons of the loudspeaker-generated ceramic drum test signal recording performed in each hall (red line) and the same dry drum test signal convolved with the experimentally derived AIRF of the corresponding hall (solid blue region). Within the bandwidth

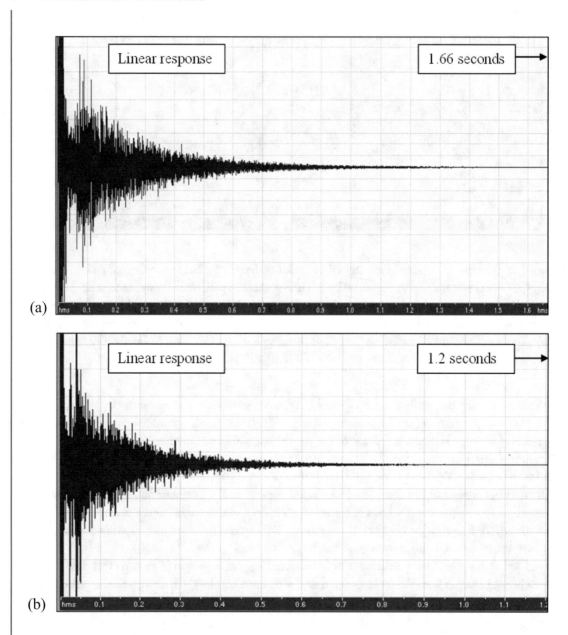

Figure 6.3: Final edited AIRF derivations for: (a) the Eckhardt-Gramatté Hall; (b) the Husky Oil Great Hall; and (c) the Rehearsal Hall. Time-scale: 0.1 s per division; amplitude-scale: 3 dB per division. *(Continues.)*

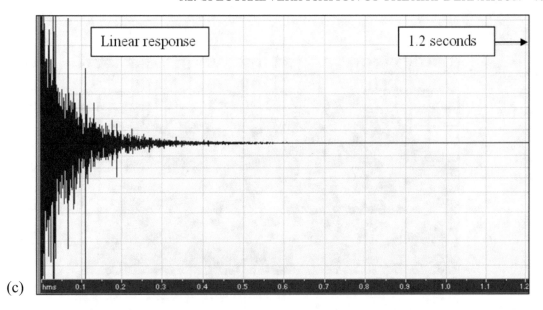

(c)

Figure 6.3: *(Continued.)* Final edited AIRF derivations for: (a) the Eckhardt-Gramatté Hall; (b) the Husky Oil Great Hall; and (c) the Rehearsal Hall. Time-scale: 0.1 s per division; amplitude-scale: 3 dB per division.

of the transducer inverse filter (100 Hz to 11 kHz) each result of convolution with the derived AIRF appears to provide a reasonably faithful representation of the drum recording performed in the actual hall.

An important parameter with the ESS method of acoustic measurement is the bandwidth of the transducer inverse filter. From each AIRF representation in Figure 6.4, the inverse filter appears to play an important role in determining the corresponding effective bandwidth of the AIRF derivation process. Chapter 4 presented a discussion on the difficulties of developing a wideband inverse filter (due to noise) and the associated problems of deconvolution. Based on the results presented in Figure 6.4, the development of a transducer inverse filter with a wider bandwidth would result in a more accurate AIRF derivation possessing a correspondingly wider frequency response.

In addition, some of the larger deviations below 100 Hz can be explained by lower frequency room modes and some issues in maintaining consistent and exact receiver positions from session to session (positions of the measurement loudspeaker source were properly maintained). Specific room modes and other acoustical considerations make the AIRF dependent upon the exact position of the loudspeaker and the microphone (an inch can make a difference) within the hall. Therefore, a difficult but necessary criterion is to maintain precise source *and* receiver locations when performing repeated measurements. This is evident from the bass frequency responses of Figure 6.4 where the ceramic

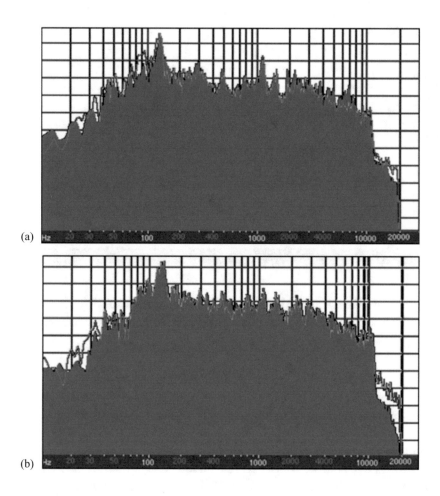

Figure 6.4: Spectra of the results of convolution of the dry, ceramic drum test signal with each derived AIRF (solid blue region): (a) the Eckhardt-Gramatté Hall; (b) the Husky Oil Great Hall; and (c) the Rehearsal Hall. Spectra of the actual hall drum signal recordings are shown by the red lines. Frequency-scale: 0 Hz to 30 kHz (log scale); amplitude-scale: 12 dB per division. Reproduced, with kind permission from IEEE, from D. Frey, V. Coelho, and R.M. Rangayyan, "Spectral verification of an experimentally derived acoustical impulse response function of a music performance hall," *IEEE Canadian Conference on Electrical and Computer Engineering (CCECE)*, Calgary, Alberta, Canada, 2–5 May, 2010, pp. 1–4. © IEEE. *(Continues.)*

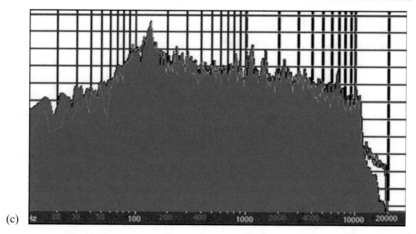

(c)

Figure 6.4: *(Continued.)* Spectra of the results of convolution of the dry, ceramic drum test signal with each derived AIRF (solid blue region): (a) the Eckhardt-Gramatté Hall; (b) the Husky Oil Great Hall; and (c) the Rehearsal Hall. Spectra of the actual hall drum signal recordings are shown by the red lines. Frequency-scale: 0 Hz to 30 kHz (log scale); amplitude-scale: 12 dB per division. Reproduced, with kind permission from IEEE, from D. Frey, V. Coelho, and R.M. Rangayyan, "Spectral verification of an experimentally derived acoustical impulse response function of a music performance hall," *IEEE Canadian Conference on Electrical and Computer Engineering (CCECE)*, Calgary, Alberta, Canada, 2–5 May, 2010, pp. 1–4. © IEEE.

drum test signal recordings performed in the actual hall were conducted in a later session from the one that produced the AIRF derivations. To maintain absolute accuracy, the ESS (AIRF) and drum signal recordings should be performed during the same session, thereby eliminating the need to reposition the loudspeaker and microphone in the exact same orientation for a future recording. Furthermore, by doing so, the accuracy of the entire AIRF derivation would be improved.

6.2.1   NUMERICAL QUANTIFICATION OF SPECTRAL DIFFERENCES

For a useful, objective analysis, it is necessary to develop a method of numerical quantification of the differences between the spectra in Figure 6.4. The MSE between two log spectra can be defined as [46]

$$MSE = \frac{1}{N} \sum_{k=1}^{N} \left[\log \frac{P1(k)}{P2(k)} \right]^2 = \frac{1}{N} \sum_{k=1}^{N} \left[\log P1(k) - \log P2(k) \right]^2,$$

where

$P1$ = the reference PSD,

$P2$ = the PSD of the simulation or derivation, and,

N = the number of sample points in each PSD.

The MSE has the important property that the minimum error of zero occurs if and only if P2 is identical to P1 [46]. This form of comparative measurement represents the residual or estimation error that exists between the reference data and the fitted model. For the present study, the reference is represented by the PSD of the drum test signal recording performed in the actual hall, while the fitted model is correspondingly represented by the PSD of the convolution of the AIRF derivation of the hall with the same dry drum test signal (see Figure 6.4).

In order to perform detailed MSE analysis, the full-length 20-second ceramic drum test signal was divided into its ten repeating segments, each having a duration of 2-s. Using the editing software in Adobe Audition, each 2-s segment was isolated separately using similar if not identical waveform editing points. For error analysis, the MSE for each segment between the two log spectra for each hall was computed using both the full bandwidth spectra of the ESS (22 Hz to 22 kHz) and the effective bandwidth spectra of the TSIF (100 Hz to 11 kHz). As evident in Figure 6.4, the bandwidth corresponding to that of the TSIF contains considerably lower error. The resulting ten-segment MSE data calculations presented in Table 6.1 support this observation. Table 6.2 provides a summary of Table 6.1.

In addition, using the same parameters as with the MSE, the integrated spectral difference (ISD) may be calculated as:

$$ISD = 10\log\left[\frac{\sum_k \{P1(k)\}^2}{\sum_k \{P1(k) - P2(k)\}^2}\right]$$

The MSE error, as presented by Makhoul [46], by itself, is a good measure, but difficult to interpret precisely. The ISD provides a measurement in dB of the spectral difference between the reference PSD, P1 (numerator), and the reference P1 minus the simulated PSD, P2 (denominator). It also provides a practical reference for the MSE calculations. Furthermore, the ISD results in Table 6.3 show that the differences in the power of the actual hall recording (P1) and the power of the difference function (P1 – P2) are approximately 30 dB over the effective bandwidth of the TSIF. This also indicates considerably lower error compared to the full bandwidth. Table 6.4 provides a summary of Table 6.3.

6.3 REMARKS

Measurement verification is important in the assessment of the derivation of the AIRF of a music performance hall. Objective numerical quantification is the ultimate goal, as listening tests only provide an aural, and therefore, subjective confirmation. Spectral analysis as illustrated in the present work provides a basic method for numerical (and visual) comparisons of a direct hall recording and convolution of the derived AIRF with the corresponding dry test signal. It is important to note the significance of the TSIF and the improvements in efficiency and accuracy that it provides. This has been a primary goal of the experiment. The ISD results of 28–29 dB over the range of the TSIF

Table 6.1: Spectral MSE calculations for each 2-s drum segment between the recorded reference and simulation data of each hall (see Figure 6.4). BW = bandwidth

Spectral MSE	Eckhardt-Gramatté Hall		Husky Oil Great Hall		Rehearsal Hall	
	Full BW	Filter BW	Full BW	Filter BW	Full BW	Filter BW
Segment 1	0.1934	0.0991	0.2057	0.1224	0.2080	0.1262
Segment 2	0.1938	0.0982	0.1962	0.1205	0.1854	0.1232
Segment 3	0.2054	0.1115	0.2216	0.1332	0.2123	0.1372
Segment 4	0.2082	0.1096	0.2066	0.1236	0.2024	0.1309
Segment 5	0.1971	0.1027	0.2033	0.1228	0.2016	0.1281
Segment 6	0.1865	0.0966	0.1893	0.1166	0.1905	0.1250
Segment 7	0.2031	0.1107	0.2194	0.1284	0.2110	0.1309
Segment 8	0.1855	0.0953	0.1885	0.1152	0.1880	0.1223
Segment 9	0.2042	0.1104	0.2091	0.1221	0.2016	0.1281
Segment 10	0.1913	0.1173	0.2005	0.1322	0.1903	0.0993
Mean	0.19685	0.10514	0.20402	0.1237	0.19911	0.12512
Standard Deviation	0.00805	0.00765	0.01112	0.00599	0.00989	0.01006

Table 6.2: Spectral MSE summary of the ten segments for each hall in Table 6.1. BW = bandwidth

Spectral MSE	Segments 1-10		Mean		Standard Deviation	
	Full BW	Filter BW	Full BW	Filter BW	Full BW	Filter BW
Eckhardt-Gramatté Hall	0.1855 to 0.2082	0.0953 to 0.1173	0.19685	0.10514	0.00805	0.00765
Husky Oil Great Hall	0.1885 to 0.2216	0.1152 to 0.1332	0.20402	0.12370	0.01112	0.00599
Rehearsal Hall	0.1854 to 0.2123	0.0993 to 0.1372	0.19911	0.12512	0.00989	0.01006

Table 6.3: Spectral ISD calculations for each 2-s drum segment between the reference and simulated data of each hall (see Figure 6.4) BW = bandwidth

Spectral ISD in dB	Eckhardt-Gramatté Hall		Husky Oil Great Hall		Rehearsal Hall	
	Full BW	Filter BW	Full BW	Filter BW	Full BW	Filter BW
Set 1	9.2885	28.8765	10.0525	29.6212	9.7309	29.2924
Set 2	9.2283	28.8231	9.5395	29.1223	8.3209	27.8782
Set 3	8.9763	28.5708	10.6612	30.2391	9.6030	29.1660
Set 4	9.1761	28.7702	9.3113	28.8858	8.2911	27.8483
Set 5	9.3736	28.7811	11.0948	30.6836	9.7571	29.3175
Set 6	9.2978	28.9054	10.0424	29.5916	9.7406	29.3217
Set 7	9.2375	28.8519	9.5490	29.1514	8.3292	27.9061
Set 8	8.9852	28.5994	9.3206	28.9147	9.5567	29.0211
Set 9	9.1669	28.7414	10.6505	30.2089	9.7473	29.2882
Set 10	9.3642	28.7523	10.1121	29.6671	9.4432	28.9934
Mean	9.2094	28.7672	10.0334	29.6086	9.2520	28.8033
Standard Deviation	0.13178	0.10448	0.58449	0.58025	0.62153	0.61607

Table 6.4: Spectral ISD summary of the ten segments for each hall in Table 6.3. BW = bandwidth

Spectral ISD in dB	Segments 1 – 10		Mean		Standard Deviation	
	Full BW	Filter BW	Full BW	Filter BW	Full BW	Filter BW
Eckhardt-Gramatté Hall	8.9763 to 9.3736	28.5708 to 28.9054	9.2094	28.7672	0.13178	0.10448
Husky Oil Great Hall	9.3113 to 11.0948	28.8858 to 30.6836	10.0334	29.6086	0.58449	0.58025
Rehearsal Hall	8.2911 to 9.7571	27.8483 to 29.3217	9.2520	28.8033	0.62153	0.61607

(see Table 6.4) are reasonable as they illustrate a noticeable improvement compared to the ISD results over the respective full bandwidth. In future work, and through the development of improved measurement conditions (such as using a real anechoic chamber for loudspeaker measurements) the goal is to achieve an ISD of 50–60 dB. In addition, the accuracy of the final AIRF derivation could be improved further with the development of a TSIF possessing a wider effective bandwidth.

CHAPTER 7

Conclusion

7.1 CONCLUDING REMARKS

The measurement transducers are still the primary limiting factor in the derivation of an AIRF of a music performance hall. In particular, the type of reference loudspeaker and its classification determine the efficiency, frequency response, and directivity with which the measurement sweep is generated and distributed throughout the hall. In the far-field environment of a concert hall, it is important that the characteristic reflections, resonances, and vibrations of the hall are properly stimulated and documented in order to develop an accurate and faithful AIRF representation. Chapter 3 presented an important discussion on loudspeaker classifications and associated physical limitations. One primary issue with any loudspeaker is efficiency. This parameter also helps to determine the level of transparency and the ability of the loudspeaker output to accurately represent the input. Due to their physical design, horn loudspeakers provide the best efficiency, and as illustrated in Chapter 3, the constant directivity beam-width of an HF horn-based loudspeaker system provides some advantages as a measurement signal generator in the far-field environment of a large hall.

For the derivation of an accurate and representative AIRF, it is necessary to prevent the external residual effects (tonal, phasing) of the measurement transducers from corrupting the final result. The present study has shown that the development and application of a proper transducer system inverse filter provides an increased level of measurement accuracy, within the specified bandwidth of the filter, in the derivation of an AIRF. The effective bandwidth of the filter is an important parameter. As presented in Section 4.4, the problems associated with noise and deconvolution make it difficult to develop an accurate TSIF with the full frequency response of 20 Hz to 20 kHz, which is essentially the same effective bandwidth of human hearing or that of a symphony orchestra. Furthermore, despite recent advances in DSP, there are still severe limitations in computing power and bandwidth, especially when processing signals of longer duration and trying to maintain a high sampling rate of 96 kHz.

Developing measuring techniques to monitor software performance (Section 2.2) is useful in that they provide additional information regarding the processing accuracy of particular modules at various stages of the acoustical measurement process. How a software product is packaged or marketed is not necessarily related to the accuracy or efficiency of the processing algorithm. A manufacturer may claim that its "convolution reverb" or measured impulse response properly represents the sound of a specific hall, but there are no commercially published data to substantiate such a claim. This is a potential problem with the current market as software manufacturers are not required to provide this type of measurement data with their product. Therefore, through measurement and analysis, the user may arrive at some conclusions regarding the performance of the software being reviewed. Specifically, frequency response analysis, as demonstrated in the present work, can easily be used to assess quantitatively the performance of a software module.

Furthermore, spectral or frequency response analysis may be used to provide measurement verification of an experimentally derived AIRF of a music performance hall. Through the use of a common test signal, Chapter 6 presented a basic methodology for numerical and visual analysis in the comparison of the derived AIRF and that of the actual hall. Measurements of the residual error or MSE between the two related spectra may be performed in order to provide a basis for objective assessment. In this manner, the present work illustrates how PSD analysis may be used for numerical quantitative analysis and verification of the measured AIRF of a music performance hall.

7.2 FUTURE WORK

The PSD analysis as presented in Chapter 6 provides a basic technique for error analysis. However, in reality, the PSD is a positive real function of a frequency variable associated with a stationary or stochastic process. A stationary process is a stochastic or random process where the statistical properties (mean, variance) do not change over time [11]. In addition, because it is a positive real function, PSD analysis does not provide information on phase. Therefore, in considering the nonstationarity of acoustical and musical waveforms, it may be necessary to examine corresponding processes of assessment such as wavelet analysis using the discrete wavelet transform (DWT). An important advantage the DWT has over the DFT is that it captures both frequency and location (time) information.

Measurement verification through numerical quantification is important in the assessment of the derivation of the AIRF of a music performance hall. Through the use of a common test signal, PSD statistical analysis as shown in the present study provides a basic method for numerical and visual comparison of a direct hall recording and convolution of the derived AIRF. The proposed method can be further improved by dividing the spectra into smaller, separate frequency bands for closer examination. This type of "microscopic" analysis will enable an increased level of numerical quantification regarding the influence of the TSIF on the final derived AIRF. This is where wavelet analysis offers an advantage in that it enables closer examination of a frequency band centered at a particular frequency [47]. In addition, specific "slices" in time of the presented audio signals may be edited and examined in the same manner in order to obtain more specific and accurate numerical information in the comparative analysis of the derived AIRF and the acoustical characteristics of the actual hall.

The methods and procedures presented in this book could lead to improved characterization of the acoustical parameters of music performance halls. A primary function of the results obtained in such measurements is the acoustical preservation of culturally sensitive and historically important halls. Furthermore, the documentation of the acoustical properties of historic architecture provides the opportunity to understand and simulate their characteristics for future endeavors.

References

[1] Gerzon, M. "Recording concert hall acoustics for posterity," *Journal of the Audio Engineering Society (JAES),* 23, 7, September, 1975, pp. 569, 571. 1, 3

[2] Farina, A. "Simultaneous measurement of impulse response and distortion with a swept-sine technique," *108th Audio Engineering Society (AES) Convention,* Paris, France, 19–22 February, 2000, pp. 1–24. 1, 5, 6, 8, 9, 10, 11, 17

[3] Farina, A., Armelloni, E., and Martignon, P. "An experimental comparative study of 20 Italian opera houses: measurement techniques," *147th Meeting of the Acoustical Society of America,* New York, NY, 24–28 May, 2004, pp. 1–22. 1, 19

[4] Bradley, J.S. "Using ISO3382 measures, and their extensions, to evaluate acoustical conditions in concert halls," *Acoustical Science and Technology,* The Acoustical Society of Japan, Tokyo, Japan, **26**, 2, 2005, pp. 170–178. DOI: 10.1250/ast.26.170 1

[5] Beranek, L.L. *Concert Halls and Opera Houses: Music, Acoustics, and Architecture,* Second edition. New York, NY: Springer, 2004. DOI: 10.1007/978-0-387-21636-2 1

[6] Moorer, J.A. "About this reverberation business," in *Foundations of Computer Music,* Eds. C. Roads and J. Strawn. Cambridge, MA: MIT Press, 1985, pp. 605–639. 2

[7] Frey, D., Coelho, V.A., and Rangayyan, R.M. "Filtering and removal of the effects of the transducers on the acoustical impulse response of concert halls," *IEEE Canadian Conference on Electrical and Computer Engineering (CCECE),* St.John's, Newfoundland and Labrador, Canada, 3–6 May, 2009, pp. 368–371. DOI: 10.1109/CCECE.2009.5090156 2, 3, 12

[8] Farina, A. "Advancements in impulse response measurements by sine sweeps," *122nd Audio Engineering Society (AES) Convention,* Vienna, Austria, 5–8 May, 2007, pp. 1–21. 8, 12, 17, 18, 21, 27, 69

[9] Frey, D., Coelho, V.A., and Rangayyan, R.M. "The loudspeaker as a measurement sweep generator for the derivation of the acoustical impulse response of a concert hall," *IEEE Canadian Conference on Electrical and Computer Engineering (CCECE),* Niagara Falls, ON, Canada, 4–7 May, 2008, pp. 301–304. DOI: 10.1109/CCECE.2008.4564543 20

[10] Fausti, P. and Farina, A. "Acoustic measurements in opera houses: comparison between different techniques and equipment," *Journal of Sound and Vibration,* **232**, 1, April, 2000, pp. 213–229. DOI: 10.1006/jsvi.1999.2694 5, 18, 50

[11] Wakerly, J.F. *Digital Design, Principles & Practices,* Third edition. Upper Saddle River, NJ: Prentice Hall, 2001. 5, 44, 84

[12] Davis, G. and Jones, R. *Yamaha Sound Reinforcement Handbook.* Milwaukee, WI: Hal Leonard Publishing Corporation, 1987. 8, 18, 19, 20, 22, 26

[13] Davis, D. and Davis, C. *Sound System Engineering.* Indianapolis, IN: Howard W. Sams & Co., Inc., 1975. 8

[14] Kinsler, L.E., Frey, A.R., Coppens, A.B., and Sanders, J.V. *Fundamentals of Acoustics,* Third edition. New York, NY: John Wiley & Sons, 1982. 8, 17

[15] Moriya, N. and Kaneda, Y. "Study of harmonic distortion on impulse response measurement with logarithmic time stretched pulse," *Acoustical Science and Technology,* **26**, 5, April, 2005, pp. 462–464. DOI: 10.1250/ast.26.462 10, 11, 27, 69

[16] Farina, A. "AURORA software suite," http://pcfarina.eng.unipr.it/aurora/home.htm. 10, 26, 46, 57, 61, 62, 69

[17] Farina, A. and Ayalon, R. "Recording concert hall acoustics for posterity," *Audio Engineering Society (AES) 24th International Conference on Multichannel Audio,* Banff, AB, 26–28 June, 2003, pp. 1–14. 17, 58

[18] ISO 3382–1 2009: Acoustics – Measurements of room acoustic parameters – Part 1: Performance spaces. 17, 58, 71

[19] Farina, A., Martignon, P., Capra, A., and Fontana, S. "Measuring impulse responses containing complete spatial information," *Illusions in Sound, 22nd Audio Engineering Society-United Kingdom (AES-UK) Conference,* Cambridge, UK, 11–12 April, 2007, pp. X1–X5. 19

[20] Beranek, L.L. *Acoustics.* New York, NY: McGraw-Hill, 1954. 19, 20, 21, 22, 24

[21] Newell, P. "Monitor systems: a background to research, part three," *Studio Sound and Broadcast Engineering,* **31**, 10, October, 1989, pp. 104–109. 19, 20

[22] Tervo, S., Pätynen, J., and Lokki, T. "Acoustic reflection path tracing using a highly directional loudspeaker," *IEEE Workshop on Applications of Signal Processing to Audio and Acoustics,* New Paltz, NY, 18–21 October, 2009, pp. 245–248. DOI: 10.1109/ASPAA.2009.5346530 20, 35

[23] Meyer MTS-4: Self-Powered Loudspeaker System, http://www.meyersound.com/products/legacy/discountinued/mts-4/index.htm 20, 28, 57

[24] Eargle, J. *Handbook of Recording Engineering,* Second edition. New York, NY: Van Nostrand Reinhold, 1992. DOI: 10.1007/978-1-4757-1129-5 21, 26, 30

[25] Colloms, M. *High Performance Loudspeakers,* Second edition. New York, NY: Wiley, 1980. 21, 24

[26] JBL 4412: Studio Monitor, `http://www.jblpro.com/pub/obsolete/4412.pdf` 23, 24

[27] JBL 4430: Bi-Radial Studio Monitor, `http://www.jblpro.com/pub/obsolete/443035.pdf` 25, 26

[28] Eargle, J. and Foreman, C. *Audio Engineering for Sound Reinforcement.* New York, NY: JBL Pro Audio Publications, 2002. 27, 69

[29] Dynaudioacoustics AIR 15: Two-Way Near-field Monitor, `http://www.dynaudioacoustics.com/Default.asp?Id=321` 29

[30] Meyer MAPP Online Pro Measurement Software, `http://www.meyersound.com/products/mapponline/pro/` 33, 34, 36

[31] Jansson, P.A. "Convolution and Related Concepts," in *Deconvolution of Images and Spectra,* Second edition. Ed. P.A. Jansson. San Diego, CA: Academic Press, 1997, pp. 1–41. 37, 41, 54

[32] Oppenheim, A.V. and Schafer, R.W. *Discrete-Time Signal Processing,* Second edition. Englewood Cliffs, NJ: Prentice-Hall, 1999. 37, 38, 39, 40, 41, 42, 45, 46, 50

[33] Denbigh, P. *System Analysis & Signal Processing.* New York, NY: Addison-Wesley, 1998. 41, 47

[34] Torger, A. and Farina, A. "Real-time partitioned convolution for ambiophonics surround sound," *IEEE Workshop on Applications of Signal Processing to Audio and Acoustics,* New Paltz, NY, 21–24 October, 2001, pp. (W2001-1)–(W2001-4). DOI: 10.1109/ASPAA.2001.969576 45, 46, 48

[35] Garcia, G. "Optimal filter partition for efficient convolution with short input/output delay," *113th Audio Engineering Society (AES) Convention,* Los Angeles, CA, 5–8 October, 2002, pp. 1–9. 46, 48

[36] Frigo, M. and Johnson, S.G. "The design and implementation of FFTW3," *Proceedings of the IEEE,* **93**, 2, February, 2005, pp. 216–231. DOI: 10.1109/JPROC.2004.840301 48

[37] Torger, A. "BruteFIR: an open-source general-purpose audio convolver," `http://www.ludd.luth.se/~torger/brutefir.html`. 48, 49

[38] Stockham, T., Cannon, T., and Ingebretsen, R. "Blind deconvolution through digital signal processing," *Proceedings of the IEEE,* **63**, 4, April, 1975, pp. 678–692. DOI: 10.1109/PROC.1975.9800 49, 51, 54

[39] Rangayyan, R.M. *Biomedical Image Analysis.* Boca Raton, FL: CRC Press, 2005. 51, 53

[40] Jansson, P.A. "Traditional Linear Deconvolution Methods," in *Deconvolution of Images and Spectra,* Second edition. Ed. P.A. Jansson. San Diego, CA: Academic Press, 1997, pp. 76–106. 52, 53

[41] Studio Projects Microphones, `http://www.studioprojectsusa.com/pdf/c_series_ manual.pdf`. 57, 58

[42] Bartlett, B. *Stereo Microphone Techniques.* Boston, MA: Focal Press, 1991. 58

[43] Sheaffer, J. and Elyashiv, E. *Implementation of Impulse Response Measurement Techniques.* Tel-Aviv, Israel: Waves Audio Ltd., April 2005. 59

[44] Beranek, L.L. *Music, Acoustics & Architecture.* New York, NY: Wiley, 1962. 71

[45] Jordan, N. "Jordan Akustik," `http://www.jordanakustik.dk`. 71

[46] Makhoul, J. "Linear prediction: a tutorial review," *Proceedings of the IEEE,* **63**, 4, April 1975, pp. 561–580. DOI: 10.1109/PROC.1975.9792 77, 78

[47] Haykin, S. and Van Veen, B. *Signals and Systems.* New York, NY: John Wiley & Sons, 1999. 84

Authors' Biographies

DOUGLAS FREY

Douglas Frey is a professional sound and electrical engineer residing in Calgary, Alberta, Canada. Through his company, *DFI Technology Ltd.,* Douglas works primarily as a sound reinforcement mixing engineer. In addition, his company provides teaching services for the Broadcast Systems Technology (BXST) and Digital Audio Certificate (DAC) programs at the Southern Alberta Institute of Technology (SAIT). Aside from his addiction to good music, Douglas is currently interested in measurement and software issues in the fields of audio and acoustics.

Douglas obtained his B.Sc. and M.Sc. degrees from the Department of Electrical and Computer Engineering, University of Calgary, Calgary, Alberta, Canada, in 2003 and 2010, respectively. Part of his M.Sc. involved recording classical Baroque lute and vocal music on location in northern Italy. His graduate thesis was largely concerned with issues of measurement integrity in the derivation of acoustical parameters of music performance halls. Douglas also has a degree in music (BMus) from the University of Calgary, where he majored in composition and orchestration.

VICTOR COELHO

Victor Coelho is Professor of Music at Boston University, a graduate of Berkeley (B.A.) and UCLA (Ph.D.), and a Fellow of Villa I Tatti, The Harvard University Center for Italian Renaissance Studies in Florence. He works primarily in the areas of sixteenth- and seventeenth-century Italian music, as well as popular music, cross-cultural and interdisciplinary studies, and media and technology. His books include *Music and Science in the Age of Galileo* (Kluwer), *The Manuscript Sources of 17th-Century Italian Lute Music* (Garland), *Performance on Lute, Guitar, and Vihuela* (Cambridge), and *The Cambridge Companion to the Guitar.* Current projects include (with Keith Polk) a history of Renaissance instrumental music, forthcoming from Cambridge University Press, and the *Cambridge Companion to the Rolling Stones.*

RANGARAJ M. RANGAYYAN

Rangaraj M. Rangayyan is a Professor with the Department of Electrical and Computer Engineering, and an Adjunct Professor of Surgery and Radiology, at the University of Calgary, Calgary, Alberta, Canada. He received the Bachelor of Engineering degree in Electronics and Communication in 1976 from the University of Mysore at the People's Education Society College of Engineering, Mandya, Karnataka, India, and the Ph.D. degree in Electrical Engineering from the Indian Institute of Science, Bangalore, Karnataka, India, in 1980. His research interests are in the areas of digital signal and image processing, biomedical signal analysis, biomedical image analysis, and computer-aided diagnosis. He has published more than 150 papers in journals and 250 papers in proceedings of conferences. His research productivity was recognized with the 1997 and 2001 Research Excellence Awards of the Department of Electrical and Computer Engineering, the 1997 Research Award of the Faculty of Engineering, and by appointment as a "University Professor" in 2003, at the University of Calgary. He is the author of two textbooks: *Biomedical Signal Analysis* (IEEE/ Wiley, 2002) and *Biomedical Image Analysis* (CRC, 2005). He has coauthored and coedited several other books, including *Color Image Processing with Biomedical Applications* (SPIE, 2011) and three on imaging and image processing for the detection of breast cancer. He was recognized by the IEEE with the award of the Third Millennium Medal in 2000, and was elected as a Fellow of the IEEE in 2001, Fellow of the Engineering Institute of Canada in 2002, Fellow of the American Institute for Medical and Biological Engineering in 2003, Fellow of SPIE: the International Society for Optical Engineering in 2003, Fellow of the Society for Imaging Informatics in Medicine in 2007, Fellow of the Canadian Medical and Biological Engineering Society in 2007, and Fellow of the Canadian Academy of Engineering in 2009. He has been awarded the Killam Resident Fellowship thrice (1998, 2002, and 2007) in support of his book-writing projects.

http://www.enel.ucalgary.ca/People/Ranga/

Printed in the United States
by Baker & Taylor Publisher Services